BEING A SCIENTIST

TOOLS FOR SCIENCE STUDENTS

MICHAEL H. SCHMIDT

BEING A SCIENTIST

TOOLS FOR SCIENCE STUDENTS

UNIVERSITY OF TORONTO PRESS

© University of Toronto Press 2020
Toronto Buffalo London
utorontopress.com

ISBN 978-1-4875-8845-8 (cloth)   ISBN 978-1-4875-8846-5 (ePub)
ISBN 978-1-4875-8844-1 (paper)   ISBN 978-1-4875-8847-2 (PDF)

Library and Archives Canada Cataloguing in Publication

Title: Being a scientist : tools for science students / Michael H. Schmidt.
Names: Schmidt, Michael H., 1962– author.
Description: Includes bibliographical references and index.
Identifiers: Canadiana 20190162791 | ISBN 9781487588458 (hardcover) |
    ISBN 9781487588441 (softcover)
Subjects: LCSH: Science—Textbooks. | LCGFT: Textbooks.
Classification: LCC Q181 .S35 2020 | DDC 500—dc23

We welcome comments and suggestions regarding any aspect of our publications —please feel free to contact us at news@utorontopress.com or visit us at utorontopress.com.

Every effort has been made to contact copyright holders; in the event of an error or omission, please notify the publisher.

University of Toronto Press acknowledges the financial assistance to its publishing program of the Canada Council for the Arts and the Ontario Arts Council, an agency of the Government of Ontario.

Canada Council   Conseil des Arts
for the Arts   du Canada

ONTARIO ARTS COUNCIL
CONSEIL DES ARTS DE L'ONTARIO
an Ontario government agency
un organisme du gouvernement de l'Ontario

Funded by the   Financé par le
Government   gouvernement
of Canada   du Canada

Canadä

CONTENTS

FIGURES AND TABLES

Figures

Tables

PREFACE

In every scientist's education, there were probably role models—teachers, authors, friends, and relatives—who made the future scientist say, "This is something I can do; this is something I *want* to do." For me those people were my Aunt Barbara; Hans Storz, a good friend's father; a whole bunch of junior-high and high-school science teachers; and, most importantly, Professors Andy Bocarsly and Maitland Jones, Jr., at Princeton, Nate Lewis at Stanford and Caltech, and Jeremy Berg at the Johns Hopkins University, who all helped start my career in chemistry. When I became a chemistry professor, I should have expected to be a role model myself, but I mostly focused on just being as good a chemistry teacher as I could be. Cards and emails I get, however, tell me that, while I'm never the *only* role model in my students' education, I have been something of a role model for students. While often this has occurred in teaching my specialty, inorganic chemistry, I have found that it also takes place in another course that I teach, one that was originally focused on just searching the chemical literature and doing scientific writing. When I was first assigned the course, I didn't anticipate that it would have a great effect on my students or me. After gradually improving the course over many years, however, I found that students were learning more than just how to look things up and how to write. They were learning, to some degree, what it means to become a scientist.

Although a lot of this book is concerned with published science research and research proposals, it is not just meant to be read by students who will someday become academic scientists publishing papers and getting federal grants. Students majoring in science can end up working in medicine, industry, start-ups, government, high-school and college teaching, or patent law. But students heading toward any of those careers must have the basic skills to understand how scientific understandings of the world come to be, how to retrieve these scientific understandings from the literature, and how to communicate these understandings to others.

Most scientists of my generation were not *formally* trained in these "softer" skills of science—the social aspects of science, the ethical aspects of science, and the conventions of scientific communication. But training took place, nonetheless: conversations took place between mentors and students, and anecdotes were told that related, in an informal way, how scientists are expected to behave, how they choose research projects, and how they get their results and theories known. Since I was trained, various initiatives have been floated by national organizations such as the American Chemical Society (ACS) and funding agencies such as the National Science Foundation (NSF) and the National Institutes of Health (NIH) to address some of these issues in a more organized fashion. The course I took over in 1996— "The Literature of Chemistry"—was started in response to a call for incorporating chemical information literacy within the chemistry curriculum, put forth by the Committee on Professional Training of the ACS. In this course, I soon found myself doing what my mentors had done—telling anecdotes that relate how scientists make decisions, how they conduct themselves in the laboratory and relate to other scientists, and how they most effectively communicate their results and interpretations. Being in charge of such a course, however, gave me both the freedom and the responsibility to organize all these anecdotes into a more systematic whole, and to do some research to ensure that I was fairly representing prevailing thought on various issues. I hope that this book will enable other teachers to more easily transmit this kind of information to the next generation of scientists.

I do not intend this to replace the training that all future scientists will get informally through conversations with their mentors, or the experience of actually doing science. Anyone who has examined the literature on the history and philosophy of science, the sociology of science, or scientific writing knows that there are no simple, fixed, right answers. My hope is that this book will serve both mentors and students in organizing the wisdom they get from official pronouncements, handed-down stories, and their own experiences.

In the course I teach, students get some experience of being an independent scientist with what I call "Household Science Projects." These were originally just designed to be a simple, safe, at-home experiment or two that provided students with something to write about, so they could practice scientific writing. But I soon realized that they were teaching students much more. I discovered that the part I took for granted—coming up with an interesting research project and seeing it through to completion—was in fact one of the hardest challenges for the students. I then had to think carefully about how scientists come up with research ideas and how to help students with this process. I also had to help students figure out where to go when their initial ideas ran into obstacles. The course that started out just being about scientific writing and literature searching gradually became a course on being a scientist. The writing and literature searching part still fits, because that's a very important part of being a scientist. But that part is now part of a much bigger whole.

I'm writing this book because I haven't found a book that presents that "bigger whole" in a way that my students can easily access. There are books on scientific writing and scientific rhetoric, books on being a creative scientist, books on scientific ethics, and books on searching the literature. However, I don't know of any books that have all of it in a compact form, written specifically for undergraduate students. That is why I have taken a little time off from the laboratory to write this book.

INTRODUCTION

THE ORGANIZATION AND USE OF THE BOOK

The chapters of this book, first written for a course I taught, were not originally presented to the students in the order in which they are given here. Other instructors may wish to present this material in a different order, and individuals may choose to read the book on their own, without me by their side to explain the idiosyncratic ordering of my course. I have therefore chosen to organize this book in a more hierarchical way than in my course. There are three parts, proceeding from the most general and abstract to the most particular and concrete: *I. Thinking, and Behaving, Like a Good Scientist*, which covers the philosophical roots of science and the ethics of science; *II. Standing on the Shoulders of Giants*, which emphasizes the skills needed to find out what is already known about a subject; and *III. Planning, Documenting, and Presenting Science*, which is about the actual process of science from the generation of ideas all the way through the presentation of finished research.

In my course, I still skip around from chapter to chapter, because what the students need in any particular week for their research projects may not fit with the more top-down organization of this book. The same may be true for individuals reading this book on their own, or students in other courses; they may need help with reading a scientific paper, coming up with a research idea, or preparing a poster before they are ready to tackle the philosophical arguments of Popper or Ziman. That's fine; I have tried to write the book in such a way that jumping between the sections is possible. The first part of chapter 1 is a good place to start, but everything students need to know for the rest of the book is covered *in just the first five sections*. The more advanced philosophy of science material, from Hume through Ziman, can be saved for later, or even left out. I have made some explicit references to other chapters within each chapter to draw connections between topics that may not at first seem related, but I have tried not to make any chapter rely on earlier chapters.

My own course, as I currently teach it, uses the chapters as shown in table I.1. In an attempt to get the students started on their research projects, I start them with the "Starting Research" chapter in week 1. I accompany this with the history and philosophy chapter, which leads to good get-to-know-each-other discussions in class. Starting in week 2, I use chapter 3 to help students begin finding good background information that they can put into proposals that they start writing in week 3. I delay the ethics chapter until week 8, because by that time I'm no longer rushing to help students get their background research done before starting on their experiments, and the students don't yet have experimental results to write about.

Other courses, which may not be organized around the household research projects as my course is, can draw readings from the book in much different sequences. Individual readers can also pick and choose at will—though, of course, I hope they eventually read the whole thing!

TO INSTRUCTORS

Instructors thinking about teaching a course like mine for the first time, or using this book to supplement a course already existing in the curriculum, may find some of the content in this book rather different from the science that they are used to teaching. Discussions, activities, and assessment concerning this content will be different,

Table I.1. How different parts and chapters of this book are currently used in CHEM 300 at California State University San Marcos.

Week	Part I	Part II	Part III
1	Chapter 1 History and Philosophy		Chapter 9 Research Ideas
2	Chapter 1 History and Philosophy	Chapter 3 The Scientific Literature	
3		Chapter 3 The Scientific Literature	Chapter 10 Developing a Proposal
4		Chapters 4 and 5 Journals and Databases	
5		Sections 6.1–6.2 Using Cited References—Backward	
6		Sections 7.1–7.2 Looking at Articles	
		Section 6.3 Using Cited References—Forward	
7		Sections 7.3–7.5 Reading for Argument	Chapter 11 The Laboratory Notebook
8	Chapter 2 Ethics		
9			Chapter 12 Writing: Grammar and Style
10			Sections 13.1–13.6 Assembling and Writing a Paper
11			Sections 13.7–13.12 Assembling and Writing a Paper
12		Chapter 8 Peer Review	
13			Chapter 14 Talks and Posters
14			Chapter 15 Final Thoughts
15			Presentations

too. I have drawn on my twenty-plus years of experience in teaching my course to write an Instructor's Manual, which is available online. Test banks, further activity suggestions, homework, and guidance on centerpiece projects are also made available to instructors online.

TO STUDENTS

Part of the standard training of a scientist takes place outside the classroom—and even outside the research laboratory. People who are more experienced in the ways of science tell less-experienced scientists stories: stories of how this research came about, how this article got published, or why that grant wasn't funded but how the results of the proposed research ended up winning somebody a Nobel Prize. You may already have heard some of these stories, but you certainly haven't heard the last of them—you will continue to hear them throughout your career. They have been, for centuries, an important way in which the unwritten rules and conventions of science have been handed down from generation to generation. Sometimes people will write a book about *some* of these stories or rules, and these are published as books on the history of science, the philosophy of science, the ethics of science, scientific writing, the literature of science, the anthropology of science, or even something like *How to Succeed in Science*. This book is an attempt to introduce you to many of the topics in those more specialized books but in a way that is well suited to an undergraduate scientist still trying to get his or her head around just what it means to be a scientist. I hope that this book puts in your brain a box labeled "On Being a Scientist"—with sections appropriately labeled as "ethics," "communication," "creativity," and so forth—so that you have a framework upon which to build as you learn from mentors and from your own experiences in becoming a scientist.

One big caution: don't regard this book as an example of scientific writing. Sure, I'm a scientist, and parts of this book are about scientific writing, but this book itself is not a scientific text—it is meant as a bridge between your friendly, casual, happy, somewhat disorganized, and not-always-properly-motivated self and the serious, thoughtful, organized, and incisive scientist you want to be in the near future. If I wrote this book using all the conventions of scientific writing, or even as a scholar of scientific rhetoric (and there are already books like that out there), I'd risk losing some of you along the way. So, don't imitate my casual and conversational style when you're doing scientific writing—instead imitate the serious examples of scientific writing that you come across in scientific journals you may read as part of this or other courses.

ACKNOWLEDGMENTS

This book would not have been possible without Steve Welch's decision to include CHEM 300 in the curriculum at California State University San Marcos. I thank him and the rest of my colleagues in the Department of Chemistry and Biochemistry for giving me the room to do whatever I thought necessary with this course, and I owe gratitude to the university, which granted me a sabbatical to make substantial progress on the manuscript during fall 2016.

I also wish to thank Steven Gimbel, professor of philosophy at Gettysburg College, for encouragement, advice, and critical reading of the first two chapters. Any errors or misjudgments in those chapters, however, are mine, not his.

I also must acknowledge Andrew Bocarsly, Nate Lewis, and Jeremy Berg, who taught me much of what I present in this book; Jeremy Berg also graciously took time from his busy schedule to review and comment on the preface, the introduction, and the chapter on peer review. Talitha Matlin and Yvonne Meulemans, librarians at CSUSM, helped me a lot with the course, and Talitha copyedited early drafts of many chapters. I also thank the anonymous reviewers for the University of Toronto Press, who gave numerous helpful suggestions, and I thank Stephen Jones, Julia Cadney, Leah Connor, and Peggy Stockdale of the Press for getting the book from an unsolicited manuscript to a published work. Margaret Allen, as copyeditor, did a fantastic job of polishing my writing, and François Trahan assisted with proofreading. I would also like to thank Rebecca Lown for the cover design, Olaf Hansen for photographs used in that design, and Shinobu Hagio and Mercedes Bonilla for use of their laboratory notebooks. Finally, I acknowledge Sergey Lobachev for his index.

Finally, I wish to thank my family and friends who have all supported me and gotten me to where I am today.

PART ONE

THINKING, AND BEHAVING, LIKE A GOOD SCIENTIST

CHAPTER ONE

WHAT DOES IT MEAN TO BE A SCIENTIST?

LEARNING OBJECTIVES

After reading this chapter, you will be able to:

- identify deductive and inductive arguments
- explain the uncertain nature of inductive arguments
- explain the importance of hypotheses
- explain that while hypotheses can be deductively proven to be false, they cannot be deductively proven to be true
- explain how methods and hypotheses are inseparable parts of what are called paradigms that define what questions are asked and what methods are accepted
- explain how science is an inherently social activity, and how this can be both a strength and a weakness

1.1 WHY BECOME A SCIENTIST?

From recollections of my own college years and my experience of teaching thousands of college students, I know that the motivations for becoming a scientist are diverse. One of the most common motivations is the promise of having skills and knowledge for which there seems to be an ever-growing demand. For some students, a science degree is a stepping-stone to a more lucrative career as a doctor, dentist, or veterinarian. There are also those who just wish to avoid the "fuzziness" of social sciences or humanities, and some who simply find that science and mathematics are what they are best at.

With all professions and academic disciplines, however, there is a process by which people with diverse initial motivations come to have a common set of goals, standards, and conventions—well-accepted ways of communicating and behaving. Sometimes this process is mostly informal, and in other cases it may be part of the formal training. In the natural sciences, it has traditionally been more of an informal process; this book is an attempt to formalize it to some extent. In the class that I teach, and in this book that I wrote for it, I have found that a more formal approach allows me to make explicit connections between the abstract philosophical goals of science and the conventions that have evolved to help scientists reach these goals.

It is for this reason that I have decided to begin with the most fundamental of questions—who are scientists, and what is science? These are questions that you may already have answered for yourself at some level. To make sure that we all start at the same level, however, I want to start with the most basic considerations of scientists as human beings, and dig a bit into the history and philosophy of science. These explorations will show us that our understanding of how scientists discover knowledge, and develop scientific theories, has continued to evolve over the centuries. Science is not just a potential career. It is a continuously evolving intellectual endeavor that is at the forefront of questions of how we understand ourselves and the world around us.

You can expect science to continue evolving in your lifetime; for example, the current conventions of how to write a journal article may change somewhat in the next thirty years. A strong grasp of the philosophical underpinnings of science should help prepare you for these changes. You will not only understand why the current conventions of scientific practice came to be, but you will also be better equipped to contribute to the conversations about how those conventions might change to bring the practice of science closer to its ultimate goals.

Although I personally think that the history and philosophy of science are interesting and useful, there are many scientists who *haven't* studied the history or philosophy of science in detail, and they still manage to be productive scientists. So, it isn't necessary for you to read this whole chapter before you proceed with the rest of this book. You can skip ahead to learn more practical tools of research and communication in parts II and III. But I would recommend that you read at least through section 1.5 for some review about the role of deductive argument and inductive argument in science. This material will help you understand other parts of the book.

1.2 SCIENTISTS ARE HUMANS

The first thing we have to realize about science is that scientists are humans. That may seem an obvious point, but it is important to remember that the strengths and weaknesses that are part of being human are also the strengths and weaknesses of scientists. Humans are highly social animals that have highly developed brains, that use language to communicate complex ideas to one another, that want to understand and explain the world they live in, and that can solve abstract puzzles. Every one of these aspects of being human is relevant to how humans do science. The last two aspects—wanting to understand the world they live in and solving abstract puzzles— are most prominent in the popular portrayal of scientists. As you read through this book, however, and as you participate in more activities that are an essential part of being a scientist, I think you'll come to see that it's also important that scientists are social animals who use language to communicate complex ideas.

The idea of scientists being a separate group of humans isn't all that old. The word *scientist* was first proposed (by William Whewell) and used in the nineteenth century; some words, like *chemist* (sixteenth century), are older, but many who today would be called scientists were, in previous centuries, just called natural philosophers. Philosophy itself is not even that old; the Ancient Greeks, who are widely thought of as the first philosophers, lived less than 3,000 years ago. That may seem like a long time ago, but agriculture is believed to have been practiced by humans for more than 10,000 years. Stone tools made by hominins—not even our species, *Homo sapiens*, but other closely related primates—date back over a million years!

So, while our idea of scientists is relatively new, some of what scientists do has been practiced by humans and their predecessors for much longer. It is hard to conceive of the earliest stone tools being developed without a pattern of trial-and-error, of a crude sort of experimentation. Likewise with agriculture: the development of agriculture not only benefited from experimentation but probably also from the recording of data. As language developed amongst *Homo sapiens*, so did the tradition of telling stories. Diverse cultural traditions have characteristic stories that are told to establish identities and to explain things that are observed in the natural world. Scientists' stories about the natural world have some particular characteristics that many of these more traditional stories lack, but a human tendency to tell stories about why things are the way they are persists in the work of scientists.

1.3 DEFINING SCIENCE, AND SCIENTISTS, MORE PRECISELY

So, yes, scientists are humans, and, for many millennia, ancient humans have been doing some of the same things scientists do: experimenting, recording data, and telling explanatory stories. In the past few hundred years, though, science has evolved into something that's quite different from general patterns of human behavior. When scientists experiment, think, and communicate today *as scientists*, they do these things differently than scientists just 150 years ago did them, and also do them quite differently from how nonscientist humans of today do them. You may think you already know how to read, write, think, and do experiments, but in my experience, undergraduate science students need a lot of practice doing these things before they can do these things as modern professional scientists would. Your writing style will have to become more direct and to the point; your reading will have to become more goal-oriented; your experiments will have to become better planned and more carefully interpreted. In some sense, you have to catch up with almost 500 years of the evolution of science.

To better understand why scientists communicate, think, and behave as they do, it is good to take a look at what philosophers and historians of science have concluded by studying the evolution of science. Sociologists, anthropologists, economists, and political scientists have joined in with their own thoughts of what science is and what it does. (Just so that I don't have to list all those kinds of people over and over again in this book, I will call them all *metascientists*—people who study how scientists do their work.) To review all of these thoughts and writings is beyond the scope of this text. A good course in the history or philosophy of science would help you better understand the historical nuances and philosophical issues in the development of science; I encourage you to take such a course, if you can fit it into your schedule. But I will try to tell a not completely inaccurate story about how *modern science* developed, starting in European cultures about 1600 CE. I think this story may help you put the more practical matters of being a scientist into perspective as you try to become a scientist.

Some people may object to the focus on Western cultures in recent times, and I understand the objection. I did just mention above that basic human characteristics of *all* humans, of *all* cultures and times, make them suited for doing the sorts of things that scientists do: observe things, invent things, solve puzzles, and tell stories about how things are and how they came to be. The Chinese developed papermaking and gunpowder, Aboriginal people in many parts of the world discovered the medicinal properties of plants, and ancient Arabic people made important advancements in algebra. But the modern science that started flowering in seventeenth-century Europe has some particular characteristics that are worth thinking about, and those characteristics are important to understand for anyone who wants to become a scientist, doctor, pharmacist, or engineer today, regardless of where they live.

It is tempting to say that science is anything that involves using the *scientific method*. You may have learned a more or less structured description of just what is involved in that method in your earlier education. But one of the things that philosophers of science—and other metascientists—pretty much agree on these days is that there isn't a single scientific method that describes what all scientists do or what they have done in the past. The best we can do is to look at the history of science and try to figure out why we trust some scientific descriptions of the world more than others. With the exception of Francis Bacon, most metascientists have generally looked backward at what seems to have worked in the past rather than prescribing what scientists should do in the future. Most of these metascientists focus on how science developed after the Scientific Revolution of the seventeenth century in Europe, and they often contrast this with what it was like in Europe before that revolution. We will take a brief look at the history and philosophy of science before and since the Scientific Revolution to try to get a better idea of what particular attitudes and methods make modern science the powerful tool that it is.

1.4 ARISTOTLE, MEDIEVAL SCHOLASTICISM, AND DEDUCTION

What came before the Renaissance and the Scientific Revolution in Europe were the Middle Ages, which are considered to have lasted roughly from the fifth to the fifteenth centuries CE. These were not quite the "dark ages" that folks sometimes portray them to be; there was in fact a lot of cultural evolution taking place. For example, the modern university is a direct descendant of the first universities founded in the twelfth century. The arts faculty at these universities, who were essentially what we would now call "arts and sciences" faculty, were heavily influenced by the writings of Aristotle. These writings had recently arrived in Western Europe in Latin translations from the original Greek, and in Latin translations of Arabic translations. Along with those writings came a lot of commentary from Greek, Jewish, and Islamic scholars who had tried to make sense of Aristotle's sometimes difficult writings. To understand what passed for natural philosophy in the Middle Ages, then, we really need to understand a bit about the "scientific methods" of Aristotle and other Ancient Greeks.

Plato and his student Aristotle were impressed by the successes of geometry in expanding human knowledge through the process of proof or demonstration. One could start with a few axioms that were evidently true and, through logic, demonstrate theorems that *had* to be true, even if they weren't immediately obvious. For example, from a few basic axioms of geometry, all sorts of interesting properties of triangles can be derived. We now call this deductive logic; we start with certain *premises* or *axioms* and end up with a certain necessary *conclusion*. If we were to apply this general idea to a non-geometrical, non-mathematical example, we might start with the following *premises*:

1. All dogs eventually die.
2. All beagles are dogs.

From this, it is a necessary *conclusion* that:

3. All beagles eventually die.

This basic structure of *deductive logic* is very important in science. If you can't get this basic sort of logic correct, you're not likely to be very successful at science. But, as we will see, the further development of science requires more than just simple deductive logic. After all, how did we get to the premise that all dogs eventually die? You may be surprised to find that people have worried about the process of getting to such a premise well into the twentieth century!

The limitations of relying on deductive logic alone can be seen by looking at some of Aristotle's arguments and conclusions about the natural world. The existence of a *void* or *vacuum* was one big problem in natural philosophy that concerned both Aristotle and the medieval philosophers. Aristotle has lots of arguments against the void, some harder to understand than others—but all are based on logical deduction from some rather abstract axioms. One that is not too hard to understand starts with the idea that a *medium* such as air or water offers a resistance to motion. This is easy to observe. If, then, there were a void, a space that was empty of any medium, there would be no resistance to motion. Aristotle also believed, however, that the medium applied a force to the object, propelling it forward. The velocity of an object was then the force applied to the object divided by the resistance, $V = F/R$. This is something which we no longer believe to be true. But with this formula, we can see that, in a vacuum, there would be no resistance, we would have to divide by zero, and therefore any object subject to any force would have an infinite velocity. On the other hand, without a medium, there was no propelling force, and without this force, there was no velocity. These two arguments suggested that an object in a void or vacuum would both have an infinite velocity and no velocity. This was an obvious contradiction, so Aristotle concluded that a void was not logical.

This conclusion was an important mistake, because one argument made against the idea of *atoms*, introduced by an earlier Greek named Democritus, was that it would require the existence of a *void* between the atoms. If you couldn't have a void, you couldn't have atoms. This turned out to be an obstacle for atomic theories all the way through the Middle Ages and beyond.

Philosophers of the Middle Ages tried to merge Aristotelean thought with the teachings of the Catholic church, and their efforts have come to be known as *Scholasticism*, because they did what scholars do—they looked at old texts and argued about what they meant. Medieval scholars did a lot of hypothetical, deductive thinking in an attempt to resolve the contradictions between Aristotle's works (which they admired for their logic) and the teachings of the church (which they had to agree with in order to keep their jobs). The biblical creation story, for example, had God creating the world in the midst of a void—but Aristotle not only said that there couldn't be a void, he also argued that the world must always have existed, and did not have a creation. To avoid these contradictions, the scholars had to generate some rather imaginative arguments, and such arguments sometimes led to new ideas and theories. Jean Buridan (1295–1363), for example, came up with the idea of an *impetus*, which was later developed by Galileo into *inertia*, and eventually led to Newton's first law of motion hundreds of years later. Once we had Newton's laws of motion, we could see the flaws in Aristotle's arguments against the vacuum, but the Scholastics didn't get that far. Although some of the ideas of these medieval scholars would prove helpful and even revolutionary, the methods of these scholars were still rooted more in speculation and deduction than in rigorous observation.

1.5 FRANCIS BACON AND INDUCTION

Western Europe saw many changes from the time of Buridan until the beginning of the Scientific Revolution. The Renaissance and the Protestant Reformation both chipped away at the authority of the Catholic church, people were less and less willing to go along with old authorities, and there was a growing sense that progress and the improvement of the human condition were possible. But Scholasticism, the merging of Aristotle's philosophy with Christian doctrine, remained at the heart of the academic pursuits, and Scholasticism wasn't making much progress in understanding the natural world. In general, Aristotle still dominated in the universities, and there was little effort to test the deductive conclusions with real observations and experience.

A lawyer by the name of Francis Bacon, however, seemed to funnel a lot of the restlessness of the age into a direct attack on Scholasticism with his *Novum Organum Scientarium*, published in 1620. The title is a reference to the *Organon*, which was the Latin name given to Aristotle's collected works on logic. Aristotle's *Organon* was the Scholastics' "instrument" of logic by which humans could learn truth; Bacon's *Novum Organum Scientarium* was the "new instrument of science," which would be a better way of making progress in humans' understanding of the world. Bacon's main point was that, while deductive logic was fine as long as it started from well-founded axioms—statements considered to be self-evident or well-established—too often the Scholastics started with axioms that weren't very well established in the first place.

Scholars assumed that the axioms of the great philosopher Aristotle must be true, and there was no reason to observe nature to establish new axioms of their own.

Bacon criticized scholars for leaping prematurely from a few poorly established axioms to very broad, general principles, and then using those principles to deduce propositions that were *less* general—what Bacon called *middle axioms*. Aristotle's arguments about the vacuum, or void, are a good example of this. On the basis of some observations of motion (which were probably not carefully quantified), Aristotle decided on very general laws of motion that we now know to be wrong. Based on these faulty laws of motion, he arrived at contradictory middle axioms about motion in a vacuum.

Bacon's proposed new instrument—the *novum organum*—was to collect many observations, and to do many experiments, and only gradually to build up to general principles. We call this method of building up *induction*. Looking back at the example I gave for a deductive argument previously, we concluded that all beagles eventually die. The first axiom or premise in this argument is that all dogs eventually die. This may seem well established, but how did it get to be well established? This premise itself is arrived at by induction, which is generalization from specific examples. Our individual premises in this case are individual instances of dogs dying, and our conclusion follows *inductively*.

1. Isaac Newton's dog died.
2. Josephine Bonaparte's dog died.
3. Richard Nixon's dog died.
4. The author's first dog died.
5. The author's second dog died.

A conclusion we could reach from these premises is that

6. All dogs die.

If we had only one example of a dog that died, our inductive conclusion would not be very strong. If we had lots of examples, our conclusion would be much stronger. If we knew of one exception, a dog that was born in 1500 BCE and is still alive, our conclusion would be weakened somewhat—although that dog could die next year. Note that we can never be sure of our conclusion here—there is always the possibility that somewhere, someday, there will be an immortal dog.

Bacon is often celebrated as the "father of the scientific method" because of his emphasis on induction. It is hard to know whether many of the great scientific advances of the seventeenth century were inspired by Bacon's *Novum Organum* or whether Bacon just captured in writing a profound change in the thinking of many people of the age. But we can see how a more observational, experimental, and inductive attitude in the seventeenth century actually corrected the very problems we found with Aristotle's thinking about the vacuum. At the very beginning of the seventeenth century, Tycho Brahe made many careful observations of the orbit of Mars. Johannes Kepler analyzed Brahe's data and found that the orbit had an elliptical shape. Toward the end of the seventeenth century, Isaac Newton developed a theoretical model that would explain this elliptical orbit, and this model included the laws of motion that we still learn today in first-year physics. These laws show that Aristotle's original understanding of motion was wrong. Independently of these developments, observations by Torricelli and Otto von Guericke established experimental evidence for the existence of a vacuum. This good inductive work overthrew many of Aristotle's faulty arguments in the century following Bacon's *Novum Organum*.

Since the time of Bacon, some philosophers and historians of science have built more careful descriptions of the methods of induction while others have protested that Bacon's inductive method is an oversimplified version of how science really works. We will examine some of these criticisms in more detail below. Nevertheless, the spirit of Bacon's criticisms is still as relevant today as it was in 1620; any working scientist can give plenty of examples of faulty arguments in the scientific literature that have been based on poorly established premises! While Bacon's *Novum Organum* does not provide a complete "scientific method," I think it captures the spirit of questioning that needs to be constantly renewed in scientific investigations. It is always worth asking, "How do I know that premise is true?"

1.6 HUME AND THE PROBLEMS WITH INDUCTION

Although Bacon's methods provided a foundation for the great progress that science has made since 1600, David Hume, an eighteenth-century philosopher, wondered just how solid knowledge based on induction could be. Hume wondered how the process of induction worked, and whether we could trust it in the same way that we trusted deductive logic. In our example of induction above, we concluded that there was a high likelihood that all dogs die based on our previous observations of dogs. But, Hume would ask us, how is our past experience of dogs any reliable predictor of what will happen with future dogs? We can only make this prediction—this induction—because those *sorts* of predictions have worked for us in the past. But that itself is an induction! We conclude that we can only trust induction by a different induction; but what justifies that induction? It, like the first induction, has some uncertainty attached. This "problem of induction" has troubled philosophers of science down through the present age.

The more radical aspect of Hume's critique of induction was that induction could be an automatic process rather than the result of deliberate thought. When you see a billiard ball rolling toward a stationary billiard ball, you know that the momentum of the first billiard ball will be transferred to the second, and the second will head off in a direction at some angle to the motion of the first. But you don't really need to think about it—you just know from experience that this will happen. A child who has watched a few games of billiards knows this will happen, without any training in physics. Some thought that Hume was essentially saying that induction was more of a psychological, rather than a logical, process. This bothered many scholars during Hume's time; people didn't want to think their conclusions about the natural world were based on a psychological phenomenon rather than pure reason.

1.7 WILLIAM WHEWELL AND HYPOTHESES

Despite Hume's doubts about induction, nineteenth-century philosophers continued to build on Bacon's inductive method. William Whewell, the same guy who invented the word "scientist," wrote in his *Novum Organon Renovatum* (renovated new instrument) that there is more to the scientific method than just collecting observations and inductively drawing conclusions. While the term *hypothesis* was in use long before Whewell, Whewell played a large role in making hypotheses an important part of what people call the scientific method. Whewell emphasized that the mere collection of many individual observations didn't automatically result in scientific laws or theories. We could, for example, observe all the dogs around us and conclude many things—that some dogs have short legs and others have long legs, for example. The example of induction that we gave in the last section required us to have a focus to our observations—we had to have a *hypothesis* of mortality. Once we knew what we were looking for, we could examine how many dogs fitted the hypothesis and from that draw a general conclusion. Whewell emphasized the importance of having some idea in mind when we make observations or do experiments.

For Whewell, a great example of the importance of hypotheses was Kepler's finding that planetary orbits were elliptical. Tycho Brahe had made careful observations of the orbit of Mars, but he didn't know what to do with them. Kepler had some ideas of what mathematical figures the orbital data might fit—some hypotheses. The ellipse was not his first choice; he tried ovoids (egg shapes), thinking that an ellipse was too simple. Eventually he tried an ellipse and found that it worked. Whewell emphasized that many hypotheses won't work, and even those that do may still have flaws. Kepler, for example, did discover the correct shape of the orbit, but he still had rather wrong notions of why a planet would take such a path around the Sun. Whewell concluded that science is not just a logical procedure of induction or deduction but requires bold invention of new ideas, many of which might turn out to be wrong.

Whewell also emphasized that a hypothesis actually allows us to proceed, at least partially, in a deductive fashion. We can use deductive logic to make predictions. If the predictions turn out not to be true, then we know, deductively, that the hypothesis couldn't be true. In chemistry, a well-accepted hypothesis of the early twentieth century was "Elements in the last column of the periodic table don't form compounds." Using deductive logic, we can predict from this hypothesis that no chemical reagent will react with xenon (Xe) to form a compound. For years, this was the experience of chemists, the hypothesis was considered a good one, and chemists called these gases "inert." In 1962, Xe was found to form compounds with fluorine under some conditions. We could then *deduce* that the original hypothesis was not true. A newer, more subtle, hypothesis had to be put forward, such as that these elements formed only a very few compounds, and only with very reactive elements. So Whewell's *hypothetico-deductive* model allowed us to get back to some of the certainty of deductive logic.

Hume's problem with induction, however, still lingers over Whewell's scheme. If our deductive predictions turn out to be wrong, we have a solid deductive basis for rejecting our hypothesis. But if we test our predictions and they turn out to be right, we are left with just another observation that agrees with our hypothesis, and so we are left with the problem of induction as outlined by Hume. We have just another dead dog, so to speak.

This is being a little unfair to Whewell, who wrote not only about simple hypotheses but about whole networks of hypotheses that together made up theories like Newtonian mechanics. For example, Newtonian mechanics, which Newton derived from planetary motions, went on to predict more than just planetary motions, and a lot of rather complex and non-obvious predictions, such as the precession of the equinoxes, that turned out to be true. Whewell's use of hypotheses, then, wasn't *just* about having another dead dog confirming the mortal dog hypothesis. Whewell would emphasize that veterinarians now have whole networks of hypotheses about how accidents, ailments, and senescence can lead to a dog's death. With these hypotheses, the likelihood of finding an immortal dog seems even more remote than it did when we relied on simply listing examples of dead dogs.

1.8 DEALING WITH DOUBTS ABOUT INDUCTION: POPPER

Karl Popper, in the twentieth century, was not too impressed by the confirmation of complex and non-obvious hypotheses that followed from a complex theory; many confirming observations and experiments still couldn't *deductively* prove a hypothesis or the theory behind it. Popper didn't think induction was a reliable way of knowing anything. On the other hand, it was obvious that science had increased our knowledge of the natural world. How did it accomplish this? Rather than trying to solve the problem of induction, Popper instead tried to solve the problem of *demarcation*. He knew that we would never get final, positive confirmation of inductive conclusions, but he wanted to *demarcate* (set boundaries around) what kinds of theories were scientific and which were not.

For Popper, a theory was not scientific because it could gain support from observations and experiments that confirmed predictions. Rather, a scientific theory was one that clearly pointed to how it could be *falsified*, or shown to be wrong. G. N. Lewis had a theory of chemical bonding that predicted the formation of compounds based on the number of outer-shell electron pairs. According to this theory, there should be no compounds of the "inert gases." When scientists found that Xe could form compounds with F, this theory was falsified. In Popper's scheme, it was a scientific theory because it could be falsified. But, because it was falsified, it wasn't a *successful* theory. So just having *falsifiable* theories isn't enough.

In Popper's view, progress in science was possible because falsified theories were replaced by better theories. Better theories were ones that were more general—that suggested more possibilities for falsification—and still managed *not* to be falsified. Popper borrowed from Darwin in writing that the theories we view as successful are the ones that prove themselves "fittest to survive."[1] For example, when it was discovered that the "inert" gases could make compounds with Xe, we could replace Lewis's simple electron-counting theory of bonding in compounds with better bonding theories based on quantum mechanics. By the time compounds of Xe were discovered, the far more general theory of quantum mechanics had made many bold and successful predictions, not only about electrons in atoms and molecules but about a host of other particles. It was a more general, more falsifiable theory because, unlike Lewis's theory, it couldn't only be falsified by certain chemical compounds. It could be falsified by all sorts of chemical and physical experiments— although so far it hasn't been. Explaining the existence of the Xe compounds using this more general, more successful theory did not take long. In fact, Linus Pauling had tentatively predicted somewhat similar compounds almost thirty years earlier based on the cruder tools of quantum mechanics that were available at the time.

As successful as quantum mechanics has been, however, it is still subject to falsification, and scientists continue to try to find situations in which it might turn out to be false. If it is falsified, new, more general theories that explain all the existing facts will need to be developed. This is probably the most important lesson modern science has taken from Popper: our accepted theories are just the ones we have decided are the best for now, and we must be prepared for the possibility that they might be falsified in the future, and that we might need new theories.

1.9 HOLISTIC VIEWS—DUHEM, KUHN, LATOUR, AND ZIMAN

As a scientist, I find Popper's writings rather abstract, without a lot of easy-to-grasp illustrations of how the notion of falsifiability is of much use to working scientists. A lot of the time I go into the laboratory thinking, "I wonder if this will work?" That question has the notion of falsifiability baked into it, but it contains a lot of other uncertainties, and falsifiability is usually not what I'm thinking about the most. Nine times out of ten, things *don't* work, but I don't claim to have falsified existing scientific theories as a result. Most of the time I just end up questioning my methods and the assumptions behind the experiment: Is this apparatus even working? Are my reagents pure? Are there other interfering phenomena going on in my experiment? In trying to answer these questions, I rely on numerous theories and methods of physics and chemistry; not only do all of these have to be correct, but my application of them to the experiment I'm trying to do has to be correct as well.

Pierre Duhem, a French physicist, historian, and philosopher of science, emphasized that in real science, hypotheses and experiments don't stand alone. Instead, they depend on whole groups of definitions, hypotheses, and theories. A single experiment does not decisively decide whether a single hypothesis is true or false; a failed prediction may, on occasion, overthrow the whole structure but more often just means that various parts of the theoretical structure need to be modified. Working scientists can't generally reduce their research to a few logical statements subject to falsification; they are at all times employing a whole network of theories and assumptions.

Thomas Kuhn was another physicist who turned to the study of the history of science. Like Duhem, he recognized that scientific hypotheses and theories could not be viewed in isolation but needed to be seen as parts of entire systems of thinking. But for Kuhn, these systems, or *paradigms*, were more than just theories and hypotheses; they were whole ways of looking at the world. These paradigms determined what questions should and could be asked, and what methods could be employed to investigate them. Transition from one of these paradigms to another involved nothing short of a revolution.

Such *paradigm shifts* were preceded, and followed, by long periods of what Kuhn called *normal science*. Normal science is largely a matter of "filling in the details" within a framework that is generally accepted. For example, physics between the time of Newton and the beginning of the twentieth century was concerned with such questions as how to apply Newton's basic laws to more complex systems or how to measure the universal gravitational constant in his theory.

Toward the end of the nineteenth century, however, the normal science of the day started uncovering observations that were hard to fit into the Newtonian paradigm. There was evidence that light had wavelike properties, but while waves in the media of air and water could be explained by Newtonian mechanics, nobody knew what medium light waves traveled through. This and other problems created what Kuhn called a *crisis* in the Newtonian paradigm. It wasn't until new paradigms emerged, such as quantum mechanics and relativity, that these new observations and understandings could all make sense again. The new paradigm did not just solve old problems but also suggested new avenues of research and new methods of doing that research.

Some scientists and philosophers didn't like Kuhn's portrayal of science very much, in part because his explanations included consideration of the psychological and sociological causes and effects of the paradigms. To some, it seemed as though Kuhn suggested that science was *subjective* (influenced by the nature of the subject who was doing the investigation), as opposed to the traditional view that science was *objective* (reflecting only the nature of the object being studied). But Kuhn argued that the values and conventions of the scientific community were such that the paradigms that scientists adopted were not random or capricious. They ultimately had to "work" in the same ways we have been discussing—there needed to be logical agreement between predictions of the theories and actual experimental observations.

Other metascientists, however, were not as careful as Kuhn in delineating the objective and subjective aspects of scientific research. They readily embraced the sociological aspect of Kuhn's ideas and employed it in a critique of science. By the mid-twentieth century, when Kuhn proposed his model of how science worked, the natural sciences were the most highly regarded academic disciplines, and lots of government money was granted to scientists to advance their studies. Being a scientist had a lot of prestige, and many people viewed science as infallible and the answer to all problems. It was therefore understandable that some metascientists emphasized the social aspects of science as a corrective to this unrealistic view.

From the late 1960s through the 1990s, sociologists and others increasingly adopted a theory of *social constructionism* that emphasizes how knowledge in a discipline is *socially constructed* through interactions between people in that discipline. While this theory doesn't explicitly say that the knowledge and concepts within a field are "less real" as a result of being socially constructed, the possibility that this might be the case hovered over a lot of contentious discussions about science in the 1990s. This was the period of what some have called the "Science Wars." In these discussions, *some* nonscientists took the work of the social constructionists to mean that science was no more authoritative or reliable than political positions or cultural traditions; on the other side, *many* scientists became defensive when they thought that their critics were denying the realities of the physical world that they had devoted their lives to studying.

Bruno Latour, one of the scholars who helped build the field of *science studies* and the ideas of social construction, seemed to put the blame for these "Science Wars" on both sides. In an article from 2002, he chided scientists for misunderstanding the point of the social constructionist critique, at least as he saw it, which was to better understand how messy assemblies of humans can in fact put together—construct—reliable descriptions of the real world.[2] In a different article from 2004, Latour's criticism is pointed at the other camp, which relied too much on a superficial understanding of the goals of the social constructionist critique, and had too little appreciation for empirical methods or the solidity of scientific facts.[3] For Latour, the social construction of scientific ideas was not a flaw but a virtue; it was a description of how people came together to decide what the best possible description of the real world could be.

From the other side of the intellectual arena, John Ziman, who started out as a theoretical solid-state physicist but whose career gradually turned toward metascience, came to much the same conclusion at about the same time. He built upon older work in the sociology of science by Robert K. Merton, who had attempted to characterize the norms, or behavioral rules, that made up what he called "the scientific ethos." Merton formulated his description of this ethos in the 1930s and 1940s—at a time when scientists in Nazi Germany and the Communist Soviet Union felt increasing pressure to have their science agree with political ideology. At that time, scientists were objecting to what they were being asked to do or to support in these totalitarian political systems. Clearly, there were norms and values that scientists held that were in conflict with what the totalitarian states of the time wanted from the scientists. Merton set out to define those norms, and came up with four: *universalism*, *"communism"* (or *communalism*, in some updated versions), *disinterestedness*, and *organized skepticism*. We'll talk about these norms more later, in the ethics chapter of this book.

Ziman took Merton's more or less sociological norms and (with a little modernization) turned them into the basis of a philosophical justification of science. In Ziman's view, the fact that science has social aspects doesn't impede the acquisition of scientific knowledge, because the norms of science channel the social activities of science in a direction that leads them away from subjective conclusions to more objective conclusions. Ziman uses the term *intersubjectivity*, a philosophical term that had been around since about 1900, to help explain why we can trust the conclusions of science. (Popper himself used the term, but he wasn't the first.) It may be true that a single scientist (or a group of scientists who are all alike in gender, nationality, and political views) is incapable of having a perfectly *objective* view of the world uncontaminated by personal or cultural bias. Intersubjectivity, however, allows the comparison and correction of potentially subjective views by comparing observations and experiments from people of different genders, different cultures, and different times in human history. Disagreements may arise, but Merton's norms work in favor of coming up with robust theories that reflect how the physical world actually behaves. Common observations and conclusions from many people of diverse backgrounds are far less likely to be the result of bias. This is a very important point, for it makes it all the more important that scientists can reliably and effectively communicate their observations and conclusions to each other—and that is why this book will devote a lot of time to talking about finding relevant scientific literature, reading that literature, and writing scientific articles of your own.

1.10 IS THERE A CONCLUSION?

At this point, you might be feeling overwhelmed by the numerous different philosophical, historical, and sociological characterizations of science. You may want to know which of these views is the "correct" one. Well, there isn't such a thing as a correct view of the process and products of science. Philosophers have been arguing about these things since before Aristotle, and they certainly aren't going to stop now. The reason I have presented these to you is that all of them are useful to keep in mind as we do science. Aristotle's deductive logic is not, by itself, enough to do science, nor is Bacon's induction the final word, but it is essential that you can recognize deductive and inductive arguments, and that you can make good ones. We have to think, like Whewell, about what our hypotheses are. We have to be skeptical, like Hume and Popper, about our inductive conclusions, recognizing that the next experiment may not fit our theories, and may demand revision of those theories. We have to be aware of how science has been criticized as being rooted in the social matrices of surrounding culture and the culture of science itself, and take those criticisms into account in refining our practice of science.

If there is a conclusion, it is that science and our view of science continue to evolve. Our current view of science is, itself, another falsifiable theory. But our current version of science, like quantum mechanics, seems to work pretty well; your main interest at this point should be to learn more about how to think, behave, and communicate like scientists do today. Much of the rest of this book is devoted to understanding the conventions of scientific thinking, behavior, and communication; I hope that, by the end of it, you will be ready to use these conventions to find your place in modern science.

FOR FURTHER STUDY AND DISCUSSION

1. Why did you decide to major in science? Think about how your decision might relate to the nature of science.
2. What *observations* might have led Aristotle to his faulty axioms about motion, and his conclusion that a void or vacuum could not exist?
3. Can you think of an induction that you typically make in everyday life that you might benefit from doubting more than you usually do?
4. What is a major hypothesis in the history of science that turned out to be false?
5. For the hypothesis you used as an answer for question 4, what new hypothesis or theory replaced the one that was falsified?
6. Imagine that a professor in a social science discipline told you that science is just another cultural belief system. In what ways would you agree or disagree with this statement?

ADDITIONAL READING

Gimbel, S., ed. *Exploring the Scientific Method. Cases and Questions* (University of Chicago Press, 2011).

    I am indebted to Steven Gimbel's work for helping me better understand and organize my thoughts on the philosophy of science, and I highly recommend his text if you want to get a taste of the actual writings of the important philosophers in this chapter.

Grant, E. *The Foundations of Modern Science in the Middle Ages. Their Religious, Institutional, and Intellectual Contexts* (Cambridge University Press, 1996).

    A fascinating look at universities, Scholasticism, and science in the Middle Ages.

Ziman, J. M. *Real Science. What It Is, and What It Means* (Cambridge University Press, 2000).

    Ziman is one metascientist in this chapter who isn't in Gimbel's book. Although this is a long, and sometimes difficult, read, there is a lot of good thought here on the changing nature of science.

CHAPTER TWO

WHAT SHOULD WE DO, AND WHY? THE QUESTIONS OF ETHICS

LEARNING OBJECTIVES

After reading this chapter, you will be able to:

- characterize moral arguments as being consequentialist, social contractarian, deontological, virtue, or ethics-of-care arguments
- apply different kinds of ethical thinking to scenarios in your present and future life and career, and think of strategies to courageously do the right thing even when it's difficult
- explain the consequences of dishonesty on the progress of science
- explain the scientific norms of *communalism, universalism, disinterestedness, originality*, and *skepticism*
- identify the different main objectives of academia, government, and industry
- recognize how the different objectives of academia, government, and industry might lead to ethical conflicts

2.1 WHY STUDY ETHICS?

We saw in the last chapter that, for metascientists like Ziman, one thing that makes science distinct from other human endeavors is how

25

scientists are expected to *behave*, and what things they are expected to do. But, if we are to examine in more detail how scientists are expected to behave, then it is good to start by considering how humans in general are expected to behave. It will be useful to see how much these sets of expectations overlap or are different. We are therefore going to devote this chapter to *ethics*, both general and scientific.

You may think you already know all about ethics. Your parents, your religious leaders, your teachers, and other authorities have been telling you what to do ever since you were young, and almost everyone will claim to be ethical. Nevertheless, it is worth thinking about what sort of ethics you have, where they come from, and what they have to do with science.

When you're young, you typically accept what older and wiser people tell you, both in science and in life. But as you get older, you will find that there are a *variety* of older and wiser people, and that they're not all saying the same things; you have to decide whom to listen to. As you grow older, you will also find that there are fewer and fewer people who are older than you, and you may realize that they aren't always that much wiser than you, either. At this point, you will have to rely more on your own wisdom and make your own decisions about what is right and what is wrong. It remains useful to listen to, and talk with, other people who have both different and similar ideas, but you won't be able to rely on other people as authorities nearly as much. That's why it's a good idea to start thinking about ethics a little more critically when you are young, because it will make the decisions you face when you're older a little easier.

In the last chapter we examined how scientists, when they started questioning the ancient authorities and started thinking for themselves, decided what is right or wrong about the natural world. This thinking was *natural philosophy*, and when natural philosophy became science, thinking about how scientists think became the *philosophy of science*. Ethics is also a branch of philosophy, but rather than being concerned with what is true about the natural world, it tries to decide what is the right way for humans to act. Essentially, ethics asks the question "What should we do?"

Of course, you're already used to thinking about the question of "What should I do?" on a minute-by-minute basis and coming up with some answers. Should I have a hamburger for lunch? How much should I study for Friday's exam? Should I talk to my friend about how much alcohol they drink? Much of the time, we don't give these questions too much thought, and a lot of the time the answers we come up with are based on achieving our own personal goals rather than the greater good of people around us and the world as a whole. But serious philosophers—and, probably, many of your friends and relatives—think that there are many instances in which you should consider more than your own wants and desires when deciding what is the right thing to do. Ethics is a matter of deciding what to do, *all things considered*. By "all things," we mean not only your own immediate happiness but also the happiness of others, the well-being of society as a whole, the health of the environment, and even something that we might call your own character.

You may wonder if deciding what to eat for lunch is really an ethical decision. There are certainly people who think that it is, and they may criticize your decision to eat hamburgers. You may have a good argument for why it's okay to eat hamburgers, or you may have just never thought about hamburgers the way they do. Ethics is neither a matter of trying to be beyond any criticism (which isn't possible) or of assuming that other people's concerns are not worth worrying about (which is about the same as saying you're always right). Rather, it is a matter of being open to thinking about what you do and why, and thinking about whether there are better and worse things to do. But there's a limit to how much you can do this; you can't spend an hour deciding what to do for the next fifteen minutes. You don't want to be paralyzed by constant uncertainty about how to proceed.

In order to avoid this kind of paralysis, philosophers and others have put together a variety of *ethical systems* that try to provide some guidance about how to make ethical decisions. The next few sections will take a look at some of the major ethical systems that can be applied to decisions in every area of human life. In later sections, we will consider the ethical norms that apply more specifically to how scientists conduct themselves *as scientists*; this includes the *scientific ethos* that we briefly mentioned in the previous chapter. Finally, we'll look at how the context in which science is done influences how scientists decide what to study and why, and how this can sometimes lead to ethical conflicts.

2.2 SYSTEMS OF ETHICS

One way to avoid being paralyzed by all the possible ethical implications of our everyday decisions is to privilege some sorts of considerations over others. Much of the literature of ethics—which stretches back millennia—consists of debates over what sorts of considerations to put over others. Some will focus more on costs and benefits, while others may focus more on the social agreements or on unchangeable rules of what is right and what is wrong. The philosophy section of your college or university's library will have many books devoted to trying to explain why different weights should be put on different considerations. Some of these books will be translations from ancient Greek, and others will have been written in the last decade. Some books on ethics, intended as introductions to the subject, will actually try to classify the different approaches of all these other books into a few main categories. That is what I will try to do here.

This is just an introduction, and I encourage you to take a real course in ethics from a real philosopher, or just to pick up some of the writings of the great philosophers and read them for yourself. If you do that, you'll find that many of the great philosophers include *many* of the themes I highlight below; few if any of them argue that there's just one simple idea that clears up all difficulties.

2.3 CONSEQUENTIALISM AND UTILITARIANISM

Consequentialism in general is the idea that we should judge the rightness of actions by what consequences they cause. *Utilitarianism* is more specific in that it tries to maximize one particular consequence: human happiness. If you think about it, many of our concerns about outcomes boil down ultimately to human happiness of some sort for somebody. Our modern everyday use of the words *utilitarianism* and *utility* has expanded beyond just human happiness; *utility* is sometimes used to mean beneficial outcomes without direct reference to human happiness. That usage may be the fault of economists. Rather than worrying about what outcomes we are concerned with, I will mostly use the more all-encompassing term *consequentialism.*

Consequentialism is the *cost and benefits* approach to making decisions. This approach to ethics is the one scientists are most comfortable talking about, because it is very results-oriented. It is an approach that is routinely employed in justifying medical experimentation. While there is a risk to patients undergoing experimental treatments, it is often assumed that the knowledge gained from the experiments will ultimately yield benefits to many more patients. These benefits will more than compensate for the cost of any harm or risk to the patients enrolled in the clinical trial.

Unfortunately, there are some difficulties in this consequentialist approach. You often can't know in advance just how much benefit will result, nor will you know exactly what the costs will be. Another problem that complicates consequentialism is that, even if we could accurately determine the utility of two or three alternatives, we can almost never consider the utility of *all* of the alternatives. It is often not only a question of whether a particular clinical trial should go forward or not but a matter of choosing between a variety of different avenues of research. Should we be spending our resources on a clinical trial of yet another antihistamine or spend it instead on searching for a treatment for malaria?

So, although it is appealing to just say you're going to do whatever maximizes some good outcome, in practice it may not be as helpful as we'd like. One thing we need to be careful of is that consequentialist arguments are sometimes just *rationalizations*—we really are deciding to do something for other reasons, and we just make ourselves feel better about the decision by enumerating the various benefits that might follow from that decision. A lot of scientists make very energetic arguments for the benefits to society that will come from their research, but these arguments can also be motivated by the benefits that the scientists anticipate for themselves if their research is funded by the government or other sources.

2.4 SOCIAL CONTRACTARIANISM

Social contractarianism is the idea that societies collectively decide what is right and what is wrong. Sometimes the contract is formal and written; in other cases, it is more implicit in social expectations for certain kinds of behavior. The idea of a social contract was very influential at the time of the American Revolution, and the Declaration of Independence reflects this. The Declaration asserts that there are certain *inalienable rights* (life, liberty, and the pursuit of happiness) that all humans have, and that when these rights are violated, the social contract has been violated. When the American colonies gained independence from Great Britain, the United States wrote a constitution, with an attached Bill of Rights, that specified further what rights, and what obligations, citizens had. Since this time, the idea of human rights has been more widely adopted and has become a common basis for ethical decision-making around the globe. An International Bill of Human Rights has been established by the United Nations and is frequently used as a means of pressuring governments around the world to work toward fairer social contracts between governments and people. Once a social contract establishes a set of rights and obligations that people have, it can be used to make ethical decisions.

Not everyone is satisfied with this approach to ethics, largely because some people dispute the idea that there are universal human rights. Sometimes, when a country is criticized for "human rights violations," the government of that country will defend itself by saying that the definitions of human rights that are being applied run counter to the cultural values of the people in the society. This has been a problem in Asia, where Singapore and China have claimed that Western ideas of human rights don't mesh well with "Asian values." Taking different cultures into account, however, might lead us to the conclusion that there are no universal rights, and therefore no universal right or wrong—a position called *cultural relativism*. If we accept cultural relativism, we have little basis for saying anything is right or wrong—couldn't we just say that the crimes of the Mafia are just part of Mafia culture? Or that female genital cutting is fine because it's part of certain cultures? Within nations that are accused of human rights violations, we usually see individuals *of those nations* claiming that their rights *are* being violated. Have they just been brainwashed by Westerners into thinking that they have rights they don't really have? Or are their governments just using "cultural differences" as an excuse not to grant citizens their rights?

If we assume that individual humans can make free and independent decisions about what is ethical and what is not, we have to accept that humans will sometimes come into conflict with the societies and cultures of which they are a part. Many of the rights that are widely recognized to be human rights in fact protect individuals who find themselves in conflict with governments, societies, or cultures: the right to free expression of dissent, the right to assemble with others of like mind, and the right to a fair trial when accused of wrongdoing. If we can't assume some of these basic rights, if we can't grant the individual the freedom to act ethically in a situation where governments or society are acting unethically, can we even begin to talk about individuals making ethical decisions? In fact, Amartya Sen and others have identified plenty of strands in Asian philosophy, literature, and history that support these same rights of freedom and dissent for individuals.[1] Social contractarians would argue that any thoughtful approach to an ethical social contract would necessarily involve some of these basic rights.

That is not to say that basing ethics on social contractarian views of "rights" avoids all problems. Many ethical questions that concern people today involve disagreements about what rights actually exist—for example, does the Second Amendment of the U.S. Constitution mean that you can always have the gun of your choice ready at hand under all circumstances? Is there such a thing as a "right to privacy," and does it extend to medical decision-making and abortion? What rights are recognized and protected is ultimately a matter of politics and negotiation, and different societies will come up with different answers. But that does not mean that the consideration of rights and obligations, established within a social contract, is an invalid approach to making ethical decisions.

2.5 DEONTOLOGY

Some people get frustrated with the difficulty of making precise utilitarian calculations for all of the expected outcomes of all possible actions, and they despair at the lack of a single, definitive answer about what rights and obligations a social contract should include. They insist that there must be some more universal, absolute laws for determining what is right and what is wrong. Ethical systems that contain such absolute laws are sometimes called *deontological* systems. *Deontology* is a relatively recent word (nineteenth century), which literally translates as *the science of duty*. Actions are judged as being ethically right or wrong depending on whether they are consistent with a central ethical duty we have as humans.

The premier deontologist, Immanuel Kant, derived one notion of this central ethical duty, but the idea of deontology is now often applied more broadly to ethical systems (including some much older than Kant's) that depend on absolute ethical rules. Some religious people have absolute ethical rules derived from divine revelation rather than from philosophical thought. Doing what God tells them to do is their duty, so ethical systems based on God-given absolute laws are sometimes also characterized as deontological systems. The command to "love thy neighbor" in the Christian religion is often cited as a general deontological rule of divine origin.

You don't need God or gods, however, to have deontological ethics; although Kant believed in God, he developed a system of ethics that does not explicitly require the existence of God or gods. Kant thought that humans shouldn't just make ethical decisions based on consequences, and he wanted social contracts to be based on something more reliable than cultural assumptions. Kant wanted people to do the right thing *regardless* of consequences or of social expectations, because he thought these were unreliable guides to ethical behavior. For Kant, *reason* was the most important human quality, and moral law had to derive from pure reason. For Kant, the absolute law was not handed down from God but derived from reason. This absolute law boiled down to a *categorical imperative* from which all other ethical laws could be derived: *Act only on that maxim whereby thou canst at the same time will that it should become a universal law*. Essentially, this says that the only valid moral laws are ones that would still make sense if everyone followed them. Or, to put it bluntly, don't do anything you wouldn't want everyone else to do!

If you really want to understand this imperative or its implications for ethical behavior you'd best read Kant himself or a more thorough synopsis than what I give here. What I hope to give you here is just a sense that pure reason might have a place in ethics, too—that sometimes what's right will not be what yields the highest utility or what society thinks is right. This is probably not an unfamiliar idea—there may well be areas of your life where you just follow absolute rules *in spite of* what you expect the consequences to be or what you think others might expect you to do. We'll see more about how such feats of moral courage might actually be performed in the real world after we look at more themes in ethical thinking.

2.6 VIRTUE

If you're like me, you'll at first find the *virtue* conception of ethics harder to understand than the previous three. Perhaps that's because *virtue* is one of those words that we use just infrequently enough to confuse with other things. Because modern people tend to think of *virtue* and *ethical conduct* as almost synonymous, it all looks rather circular—how can we use virtue as a guide to what is ethical, if we define virtue as ethical conduct?

Part of the reason for this confusion is the age of this idea. The last three ideas—consequentialism, social contractarianism, and deontology—had their strongest advocates in Europe in the last 300 years, during the Age of Enlightenment, when intellectuals focused on understanding everything as rationally as possible and providing a logical framework for making decisions. Some people have characterized these approaches as *analytical*; these approaches seem to promise that an objective *analysis* of an ethical dilemma can be made that determines the proper course of action.

The concept of virtue is usually traced back to Aristotle, and although much of Aristotle's philosophy is rather analytical, his writings on ethics are less so. If we want a better feel for what this virtue thing is, it pays to look at what Aristotle wrote. *Virtue* is how most modern translators translate the Greek *arete*. But, according to some translators, the meaning of *arete* might be closer in some way to excellence or flourishing. For Aristotle, human virtue is about being the best human you can be.

Aristotle's virtues are less precise than Kant's categorical imperative. They are mostly about finding the right middle—the middle of courage between cowardice and recklessness, the middle of moderation between asceticism and decadence, or the middle of generosity between stinginess and indulgence. I'm well aware that a lot of those words in the last sentence aren't ones we use every day—I thought about replacing them with more common ones, but I think the translators from whom I borrowed them chose them for a good reason.[2] Other, simpler, more common words are a little different—they emphasize what we in the twenty-first century care about and aren't quite faithful to what Aristotle is describing.

I think the take-away message from Aristotle is not the particular virtues he described but the notion of virtue and the way of achieving it. For Aristotle, ethics isn't about having the right formulation of logical principles ready at hand when an ethical decision must be made. Rather, it is about *being* virtuous, about having good *character*. Aristotle emphasized that virtues could only be acquired by continuous practice. You can't just be courageous when a big ethical challenge comes up; you have to cultivate courage by being courageous in your everyday life. If you have cultivated virtue in yourself, then you are more likely to do the right thing when the big challenges arise.

2.7 ETHICS OF CARE

Another more recent approach to ethics that is different from the more analytical approaches of the Enlightenment is the ethics of care. This approach to ethics came to prominence when a female psychologist, Carol Gilligan, found that the way in which women talked about ethical decisions focused more on context and relationships than on abstract principles. While some philosophers and psychologists had previously interpreted these sorts of results to mean that women were less sophisticated in their ethical reasoning, Gilligan suggested that the more contextualized and relationship-based reasoning typical of many women might in fact represent an equally valid approach to ethical questions. She and others raised the possibility that traditional approaches to questions of ethics were biased toward the way men naturally thought. After all, the philosophers who had developed consequentialism, social contractarianism, deontology, and virtue were mostly men.

In the last few decades, a number of philosophers have further developed the concept of an ethics of care that emphasizes the relational and contextual thinking noted by Gilligan. The analytical approaches to ethics that arose in the Enlightenment are built around the idea of independent, equal, self-interested people interacting with others who are very often strangers to each other. The ethics of care, however, emphasizes the fact that all people start life as very *dependent* people—children—being cared for by people who are *not* strangers; throughout life, these people will find themselves at times caring for others and being cared for by others, as when they are sick or suffer the effects of old age. To the philosophers advocating the ethics of care, any system of ethics that doesn't consider the caring relations that are central to human experience is necessarily going to leave out a lot of necessary and good behavior that needs to be included in a consideration of what is ethical.

One thing that makes the ethics of care a bit different from the more analytical schemes of ethics is that, although the importance of care is a central idea in determining what is ethical, it is not assumed to be a single dominant concept like the idea of utility in consequentialist ethics or the categorical imperative in Kant's deontology. Advocates of the ethics of care admit that consequentialist, contractarian, and deontological arguments still have a place in thinking about ethics; their main point is that, without a sense of care, legalistic application of these ideas could lead to morally undesirable outcomes. Concepts of justice and human rights are good vehicles for getting societies and even global communities to a more ethical place, but the idea of caring relations between humans can be a moral compass helping to direct such travel.

2.8 USING DIFFERENT APPROACHES TO ETHICS

So, we have all these different approaches to ethics. What do we do with them? Do we have to decide on one or another? No. Most people, even those who have a very philosophical view of ethics, tend to use a combination of these different approaches in making everyday decisions. The reason we identify and classify these approaches is not to make you choose a single one but to make you more thoughtful about the reasoning you are using in making your decisions. Are you really only considering the consequences of your actions, or are you also fulfilling an implicit social contract? Are you making simple deontological rules for yourself like "always tell the truth" rather than trying to figure out when telling little "white lies" might be okay within the rules of the social contract? Are you primarily interested in making yourself the best you can be by cultivating virtues? Are you considering how a caring person would act within the context of a caring relationship?

Once you've identified what approach you're using, you're in a better position to be critical of your decision and to think more clearly about other options. Once again, you don't want to think yourself into paralysis, especially for the little decisions. But for the big decisions, consciously thinking about the different ways of thinking about those decisions may well lead to a calmer, more confident decision. Sometimes you'll reject actions based on a social contract because you don't think that the standards of conduct written into the contract are ethical enough. In such a case it's good to know why and how you came to that conclusion, whether by thinking of consequences, a categorical imperative, or a caring relationship. At other times you may reject a simple deontological rule because the consequentialist in you is alarmed at the harm that would result. Knowing that different thoughtful and intelligent philosophers over the ages have had good things to say about all these approaches should enable you to feel confident that they are reasonable and valid approaches and that it might be good to consider multiple approaches when you are unsure of what to do.

2.9 ETHICS IN PRACTICE

In teaching ethics, we sometimes present students with case studies that introduce challenging ethical decisions. Usually these are the sorts of decisions where different people using different philosophical approaches or different values can justify a number of different answers. These sorts of problems are great for having passionate discussions, and they may help raise the awareness of important ethical issues of the day, or get people thinking in different ways about these issues.

But do these sorts of discussions make people more ethical? In some cases, they might, if they make people realize all the different issues and viewpoints that need to be considered. Often, however, the difference between an ethical person and an unethical person is not that the ethical person has a deep understanding of different ethical systems and the unethical person doesn't. Nor is it that the ethical person has spent more time considering really challenging ethical situations. It many cases, the unethical people *know* what the ethical thing to do is, but they don't *do* the ethical thing. Why don't they?

In some cases, it may be that the unethical people anticipate exorbitant personal or psychological costs for doing the right thing. Let's imagine a scientist at a pharmaceutical company. Perhaps he or she will be thinking along these lines: "If I don't fake the data, the company won't be able to market this drug; my branch of the company will lose funding; I'll lose my job; my family will have to move; I'll have to take a job that pays less, has less prestige, and doesn't give me the freedom to explore the science I want to explore; and I'll be unhappy."

If that's what someone is thinking, it isn't too hard to imagine that person making some unethical decisions. What is the best way to deal with such depressing thoughts? I would recommend thinking through the following considerations:

1. Critically examine the "chain of causation" you have in your head. A lot of times we tend to think of the worst-case scenarios. Things may not be that bad—maybe the first few steps will happen—the drug won't get marketed—but will that necessarily lead to you losing your job? Maybe the company has another drug that will get through clinical trials and make the company

rich. Maybe you will lose your job, but you'll find a better job at a different company. Maybe you'll end up teaching high school chemistry and find that you love it more than working for that pharmaceutical company!

Use your quantitative reasoning powers to critically examine your chain of causation. Long chains of causation have low probabilities of making it all the way to the end—a fact that is well understood by those doing multi-step syntheses of complex organic molecules. Maybe there's a 75 percent chance the drug won't get marketed if you don't fake the data, a 75 percent chance that the company loses money, a 75 percent chance that you lose your job, a 75 percent chance that you'll have to take a job that pays less, and a 75 percent chance that you'll be less happy as a result. What are the chances of all these happening? $(0.75)^5 = 24$ percent! The worst-case scenario has less than a 25 percent chance of happening, despite pretty good odds that any one of these bad things will happen.

2. Don't just think about the negative consequences of ethical behavior, but also think about the consequences of the unethical behavior that you might be tempted to engage in. What if you lose your job because somebody found out you faked the data? Your chances of getting another good job are a lot worse if you have a reputation as an unethical scientist than if you *honestly* failed to produce the data people were hoping for. There are lots of people who would rather hire an honest person with only a few minor successes than a highly successful criminal.

Your ethical lapses tend to be regarded more seriously when you're working in science than in some other occupations. Scientists in general have pretty high standards and pay attention to details. There probably haven't been too many people in advertising who are unemployable because they were found to have stretched the truth a bit on occasion, but there have been scientists who have had to essentially quit doing science because they weren't scrupulously honest.

The possible negative consequences of unethical behavior don't just apply to individuals. In the summer of 2012, GlaxoSmithKline was fined *3 billion dollars* for using false information in marketing some of its most popular drugs. It may seem like an old and trite saying, but "honesty is the best policy" really does make sense in science and medicine.

Now, the previous two suggestions will take care of many normal, everyday temptations to do the wrong thing. But sometimes the problems will be harder. Sometimes the situation is that there's a 99 percent chance you'll lose your job unless you do something wrong, while there's a 99 percent chance that you'll not suffer any negative consequences for doing the wrong thing. This sort of thing might happen in industries where breaking the law or other moral codes is a well-accepted part of doing business.

If that's the case, you may sometimes find yourself at a point where it really does look like doing the ethically wrong thing is going to be a better option for your own happiness. If your system of ethics demands that other considerations be put above your own happiness, you have to rely on *moral courage*. People with real moral courage do the right thing *despite* the very real prospect of unfortunate consequences.

2.10 ABOUT MORAL COURAGE

Moral courage isn't easy, but there are some steps one can take to make it easier. One step is to keep in mind the moral reasons for your decision. If you came to your ethical decision from a utilitarian perspective, focus on the ultimate human happiness it will bring to society as a whole when you do the right thing. Preventing a useless or harmful drug from making it to market may not benefit the company or the employees of that company, but your focus should be on all the thousands or millions of people who end up paying for a useless drug or suffering side effects that aren't necessary.

Another step that you can take is to practice moral courage. I think Aristotle was right in emphasizing the need to *practice* virtue, and courage is one of his virtues. People who are used to doing the right thing in spite of possible costs or inconveniences are more likely to do the right thing when really courageous action must be taken.

Fortunately, we have plenty of opportunities to practice moral courage. Not all ethical choices involve anything as dramatic as risking your livelihood. A lot of times the fear that drives us away from ethical action is nothing more than the fear that people might look at us funny for worrying about something that nobody else is worried about. These smaller challenges, however, are just the sort of situations that allow us to develop habits of ethical behavior—to develop virtue. Practice doing the right thing even if it's a little thing. If you mess up, try to admit your error and fix your mistake. Practice adds to your ethical self-confidence and your moral courage.

A final step you can take to make moral courage easier is to surround yourself with examples of moral courage that can serve as models for you when things get tough. Read books and watch movies about heroic people who do the right thing. Talk with your family, friends, and colleagues about how you would support them, or how they would support you, when doing the right thing brings risks to our safety and security. If you don't know people who share your moral values, find some! You are currently spending much of your time preparing for your future career by studying science and mathematics many hours per week—wouldn't it be good to spend a little time preparing for your moral future as well?

2.11 THE ETHICS OF SCIENCE

When people talk about "scientific ethics," there are at least two different categories of questions they might be talking about. One category of questions is about whether the actions of scientists conflict with the ethics of humanity as a whole. Under this category, we have questions like whether it is ethical for scientists to permanently edit the genes of humans, or what the limits might be to doing experiments on our fellow human beings. For questions like this, we really have to go back to our "ethics in general" approaches and simply apply them to situations in which scientists find themselves. That's an important thing to do, and it's one reason we covered ethics in general before we covered anything else. We could spend a whole semester on such issues—many people offer courses, and even degrees, in scientific and biomedical ethics involving these sorts of questions. But that's not the category of "scientific ethics" that we will consider in this section.

Another category of "scientific ethics," the one that we *do* want to consider here, is more concerned with questions like "What rules do scientists generally follow, and expect other scientists to follow?" To avoid confusion, we can refer to these as the "*norms* of science." Some people might insist that this is more a question of sociology than ethics, and in fact we will end up looking at what sociologists have found about the norms, or rules, that scientists follow. But I don't want to draw a sharp line between the (mostly) philosophical view of ethics in general of the last few pages and the more sociological view of the scientific *ethos* of the next few pages. We will still find common ideas, because ethics in general and scientific norms both involve getting people to work together productively and establish trust within a community.

In the previous chapter, I introduced some of the thoughts of John Ziman, who started as a physicist but ultimately became a *metascientist*, someone who studies how science is done. In Ziman's view, the sociology and philosophy of science can't be cleanly separated—the social realities of how science is done end up affecting how things about the natural world get to be known, and the philosophical reasoning about how to properly know things about the natural world in turn affects the social realities of how science is done. We're going to take a Zimanistic approach to scientific ethics and consider both the philosophical and sociological aspects of the rules scientists follow in doing science.

2.12 THE IMPORTANCE OF HONESTY

One thing we learned from our initial chapter on the philosophy of science was that induction is an important method in the development of scientific ideas. Scientists generalize from lots of observations to more general conclusions. An important aspect of the inductive process is that we benefit from being able to trust the observations of others. If we can't trust the reports of other scientists now and throughout the history of science, we don't have as many observations to go on, and our conclusions aren't as strong. Scientific progress is made possible by trust; if you can trust what other scientists have reported, you don't have to start from scratch and do every experiment ever done before.

The whole system falls apart if people aren't honest. Therefore, I would say that the number one ethical rule for scientists is: *be honest*. Be scrupulously, exactingly, thoroughly honest at all times and in all cases. Honesty is part of the *social contract* among scientists. You could also think of the rule of honesty as a *deontological* rule; it is your duty as a scientist. If you prefer to think in terms of *virtue*, cultivate the virtue and habit of honesty as the most important virtue for a scientist. And the *consequence* of honesty in the scientific context should be clear—it allows us to trust the store of observations of scientists throughout history, resulting in theories and understandings that enable new technologies that add to the happiness—the *utility*—of all humanity. If you *care* about humanity, and the scientists who will build on your work, you won't want to mislead people.

Honesty is a *general* ethical rule, not one that should only appear in a list of rules for the practice of science. But scientists probably emphasize honesty more than other groups of people. There are other professions in which honesty is not as highly valued—you might have your own ideas of which these are. There are philosophers who argue about whether white lies are acceptable or even desirable in nonscientific contexts. You may even find yourself being less than scrupulously honest in parts of your life that aren't related to your scientific endeavors; it might be okay to tell your boyfriend that his new haircut "looks great." But in the realm of science you must be scrupulously honest. So, while honesty is valued outside of science, I think the value placed on honesty is greater within science, and the expectations of honesty are stricter.

One important aspect of the honesty that scientists must have is the ability to be honest with themselves. It is not always easy to say to yourself, "I don't know," or, "I'm not sure." But very often that is the honest truth, and it is an honest truth that is very important at the beginning, the middle, and even at the end of a lot of research projects. It is the sort of very honest thing people must learn how to say to themselves in a lot of different situations. We will see later that this honesty leads to a careful writing style that avoids bold statements about what is "known" or "proven."

One very good explanation of the sort of honesty required in science is by the late physicist Richard Feynman. There is an essay of his called "Cargo Cult Science," adapted from a 1974 commencement address he gave at Caltech. It can be found online if you look for it; it's also published in his book *"Surely You're Joking, Mr. Feynman!"*[3] The references in the essay—and some of the cultural attitudes Feynman reflects—may seem to the modern reader to be somewhat sexist and/or colonialist, but the piece captures much of the spirit of Bacon's *Novum Organum* in a way that is easier for twenty-first-century people to read. It is about what Feynman calls "a kind of scientific integrity, a principle of scientific thought that corresponds to a kind of utter honesty—a kind of leaning over backwards": that is, leaning over backward in the sense of not just saying honestly what is obviously true about your research but also trying to honestly, and imaginatively, examine all the ways in which you might be wrong. Trying to imagine how you might be wrong—and how to prove that you are wrong—is the essence of Popper's idea of falsification, discussed in chapter 1.

2.13 THE ETHOS OF SCIENCE

One of the most widely accepted formulations of what scientists think they should be doing is Robert K. Merton's *ethos of science*, which I briefly introduced in the previous chapter. Merton formulated his description of this ethos in the 1930s and 1940s—at a time when scientists in Nazi Germany and the Communist Soviet Union felt increasing pressure to have their science agree with political ideology. Scientists were objecting to what they were being asked to do or to support in these totalitarian political systems. It appeared that scientists had norms and values that were in conflict with what the totalitarian states of the time wanted from the scientists. Merton set out to define those norms, and came up with four: *"communism,"* *universalism, disinterestedness,* and *organized skepticism.*[4]

We have to be a little careful in looking at them today—science has not necessarily stayed stagnant over the last 80 years, and the pressures being put on scientists are different than they were then. Nevertheless, there still are situations where some governments lean on scientists to come to acceptable conclusions, especially in such areas as environmental issues. These days, the pressures on scientists tend to be more focused on making science profitable or making it less likely to interfere with the profitability of some businesses.

As old and as history-bound as Merton's norms may seem, however, a lot of people who have written about the history, philosophy, and sociology of science in the last few decades still find them relevant, though they have made some changes. John Ziman's adaptation of Merton's norms is widely accepted today as an updated version of those norms. In describing the norms that define the ethos of science below, I have depended more heavily on Ziman's characterizations and ordering of these norms than on the original formulations of Merton. Ziman's descriptions, published in 2000, have the advantage of taking into account a lot of experience and debate since Merton first put the norms down on paper.[5] Here are the norms, in the order, and with the names, used by Ziman.

Communalism (Merton originally had it as "communism," in quotation marks): This is the idea that knowledge is held communally amongst all scientists. It means that information is shared, promptly and in full. Knowledge isn't kept private. From a consequentialist point of view, the good thing about this norm is that it encourages scientific progress; scientists can build on each other's results only if they know what these results are.

It is worth wondering if this is really a norm of science today. More and more, science is done by private companies who are interested in *owning* the knowledge they generate, so that they can profit from it. Myriad Genetics claimed to *own* the gene sequences for BRCA1 and BRCA2, two genes in which mutations may increase the risk of breast cancer. Others challenged this ownership, all the way to the Supreme Court of the United States, and this claim was partly thrown out. But there is a lot of research in technical fields that is legitimately owned through patents, both by private companies and other organizations. A lot of universities have made a lot of money from patents they hold; Princeton University financed a fancy new chemistry building from the profits earned on the patent on the anticancer drug Alimta (pemetrexed) developed by Professor E. C. Taylor. So, is this norm of *communalism* really a valid reflection of what characterizes science today, with all this privately held, patented, profitable science?

I would say the answer is a qualified "yes." The two examples of patents I gave might clarify what is kept private and what is shared. The purpose of patents is to reserve the right to *profit* from inventions or discoveries for a given period of time, while allowing information about the inventions or discoveries to be published. The information is often shared freely with the scientific community through the published literature. The scientists at Myriad Genetics didn't keep the BRCA1 gene a secret; they published an article in *Science* on it that included the amino acid sequence coded for by the gene.[6] Many articles followed this first article, describing further research on this gene and its product, even as Myriad was developing clinical tests that used the information, and from which they hoped to make a profit. Likewise, E. C. Taylor, together with other scientists, promptly published news of the synthesis and initial indications of antitumor activity for the molecule that became the drug Alimta. This was published some years before clinical trials were complete and the usefulness of the drug in patients was established.[7]

The communal sharing of information, then, needn't be opposed to the capitalism—the money making—that science often serves. The norm of communalism is really about sharing *information*, even if the rights to profit from this information are held privately. It would actually be much *harder* to profit from knowledge of new pharmaceutical molecules or disease-linked gene sequences if the information *wasn't* shared through publication. Nobody would pay good money for Alimta if evidence for its efficacy were kept secret. People wouldn't pay for a genetic test for mutations in BRCA1 if scientists worldwide weren't able to see the evidence that such mutations contributed to breast cancer risk. Science, even of the profit-making variety, doesn't work if information isn't shared.

I feel obliged to point out, however, that this interpretation of communalism is a little subtler than Merton's original description of the norm he called "communism." Merton actually directly contradicts what I argued in the last paragraph: "The communism of the scientific ethos is incompatible with the definition of technology as 'private property' in a capitalistic economy."[8] Merton saw the issuance of patents as in direct conflict with the scientific ethos. There are scientists around today who feel the same way, but there are also thousands of people working in private, for-profit, research labs to develop patentable molecules or instruments who would consider themselves scientists. You may become one of them! I therefore think it's a little unrealistic to consider Merton's idea of communism as really descriptive of the entire scientific community today.

Of Merton's original norms, communalism is probably the one that has changed the most since Merton first wrote of the norms, and it may continue to change in the future as scientists continue to balance the need for openness with institutional survival. Advanced science and technology can be expensive, and having mechanisms in place to pay for it will mean that a science free from any notion of ownership is unlikely. In the middle of the twentieth century, sufficient government funding of science to make all scientific information freely available for use by all seemed possible to some people, but without large political changes, that scenario looks increasingly unlikely.

Universalism: This really just means that the scientific information that is shared isn't given different status depending on who shares it. Who you are doesn't—or at least *shouldn't*—matter. The "universal" in "universalism" is really a matter of universal opportunity for scientists to share their work and discuss the work of others, regardless of race, gender, socioeconomic status, or place of origin.

We have to keep in mind that these norms are, in a sense, ideals; in reality, scientists are more likely to cast a critical eye on the work of a young scientist from Tanzania than on the work of a senior faculty member at Caltech. But the norm is there, and scientists are aware of it. If a scientist's criticism of another scientist is based primarily on who the scientist is, or where he or she comes from, the criticism is more likely to be disregarded, and the reputation of the criticizing scientist may be hurt as a consequence of violating the norm of universalism. Scientists are not free from prejudice, but the norm of universalism requires that they strive to make their judgments of scientific work free from prejudice.

The result of this norm is that criticisms of the work of other scientists are usually based on actual problems with the work, not on who did the work. Criticism is still important; we will see more about this under the norm of skepticism, discussed below. But the norm of universalism encourages scientists to avoid *ad hominem* arguments, one of the *logical fallacies of irrelevance* in which the person, rather than the argument presented by the person, is attacked. If you've taken a critical thinking course, you may recognize this fallacy as a common mistake in discourse or discussion that leads to illogical, and ineffective, arguments. The norm of universalism thus has beneficial consequences for the science itself, helping scientists avoid errors that could slow the progress or lower the quality of science. Like the norms of honesty and communalism, it is a norm that has definite utilitarian consequences.

Disinterestedness (and humility): As with communalism, we have to be careful about interpreting this word. *Disinterested* means *not having an interest*, as in *not having a personal or financial interest*. But scientists almost always do have something of a personal or financial interest in their science, even if they are at an academic institution. People get promotions and raises based on the success of their science. So how can any scientist be disinterested?

As with universalism, disinterestedness is more about how the communication of scientific ideas takes place than about what actually motivates scientists on a psychological level. Certainly, we have to be aware of the financial and personal interests that we, and other scientists, have. There are an increasing number of policies in place that require scientists to disclose their financial interests in their scientific communications. But it is still possible for scientists to present their work with a *rhetorical attitude* of disinterestedness. In other words, scientists have to demonstrate with their communications that their conclusions rely on experimental fact and logic rather than on what they want to be true.

One manifestation of disinterestedness is the style of scientific writing. It's frequently in the passive voice and filled with references to other people's work. We'll discuss more about this in a later chapter where we examine scientific writing in detail; what it means for us now is that the writing is designed to convince others that it doesn't matter who is doing the writing, or who did the actual experimental work, or what motivated the writer or experimentalist. What matters is the experiment, or the interpretation, or the theory itself. The author is meant to be as invisible as possible.

Ziman connects this rhetorical practice of disinterestedness with a general stance of *humility*. Citing other people's work acknowledges the scientist's debt to other scientists and also shows that the work is not dependent on just one person's ideas and motivations. It reinforces the communal nature of science and deemphasizes the importance of a single personality in coming up with good science.

Ziman also sees the norm of disinterestedness as a way of encouraging honesty. If a scientist really had a financial or egotistic interest in seeing his or her theories proven true, could a scientist actually *lie* about what experiments were done or what the results showed? We unfortunately have plenty of examples in the history of science in which scientists did exactly that.[9] In some cases it wasn't just a matter of lying to others but of lying to oneself. *Self-deception* can be a problem in science, and it's most likely to happen when a scientist has an especially strong interest (in a financial, social, or egotistical sense) in the results. We therefore may benefit from not just having a rhetorical disinterestedness but trying to have a bit more of a genuine psychological disinterestedness. This could help us achieve the important goal of being more honest with ourselves, which I mentioned in the previous section.

It works the other way as well—attempts to be more honest with ourselves could also push us to greater disinterestedness. For example, let's be honest with ourselves: will faking experimental results really help us achieve fame and fortune? No, our chances of being caught are pretty good, and the consequences can be pretty devastating. It's better, then, to be more interested in the universal, communal goal of scientific progress than in our own desires for fame and fortune. Scientific progress will always benefit from honest exploration, regardless of whether the results work toward our financial, egotistical, or social interests.

Originality: This norm wasn't present in Merton's original list, and it's a little different in tone and effect, too, from the norms in Merton's list.[10] The previous three norms all push us in the direction of making scientific results and theories more likely to stand the test of time. If we openly share results with one another (communalism); if we judge results and ideas based on logic and evidence alone, without regard for who shares them (universalism); and if we discuss results without regard to what our personal, financial, or social interests are (disinterestedness), then the scientific theories and laws we come up with are more likely to be good reflections of what nature is really like. But these norms do not push us forward to expand these theories and laws into new areas. The norm of originality is what pushes us forward. This means that *new* observations, experiments, and theories are the ones that are the most valued.

All of Merton's original norms were identified by the sociologist as norms of the scientific community, but they were the sorts of norms that correlate well with the norms of many ethical systems outside the scientific community. Sharing, not being prejudiced, and not being selfish—these are the sorts of things people all over the world learn from their parents, their religious leaders, and their teachers. But the norm of originality is a less universal norm, and as such it really is something we *observe* in working scientists, as a sociologist would. It's not unique to science—communities of artists and scholars in nonscientific disciplines also value originality. But it's by no means a universal norm shared by most people, and it doesn't clearly follow from a lot of the ethical systems we considered in the first part of the chapter.

Nevertheless, it is not without ethical implications. "Progress" is usually seen as a good thing, as it results in better nutrition and medicine and more comfortable built environments. You can (and I have) gotten into lengthy online arguments about just how morally important progress is, but it is sufficient to recognize here that a pretty good case (mostly consequentialist or utilitarian) can be made for scientific and technological progress as having ethical implications. Without originality, there is no progress. So, while many ethical systems outside science don't emphasize originality, it does have a greater moral dimension to it.

Ziman points out one possibly negative side effect of this emphasis on originality: greater specialization. In order for scientific results to get published, they have to be original. For this reason, the literature-searching skills that will be emphasized later in this book are important. You need to know what has already been done in a research field, not only to make your own research easier but also so that you don't present your work as new when similar or identical work has already been published by someone else. Finding new research problems that haven't been worked on before is hard; there have been many, many scientists working for many centuries on many different problems. As a result, most scientists become very specialized in one narrow field, as it is the best way to make sure that they have a good idea of what has been done before, and what hasn't. This drive toward greater specialization *can* have the disadvantage of discouraging bold ideas that span across different specializations.

Another possible disadvantage of an emphasis on originality is that it doesn't encourage scientists to redo other scientists' work to confirm that it is repeatable and correct. It's very difficult to get published if all you do is repeat other people's experiments, unless you can show that previous experiments were wrong. Because publication is a major measure of scientific success, there's not much reward in just trying to see whether other people's research is correct. But a lot of folks these days are worried about the fact that too few people follow up on promising results to make sure they are repeatable and correct. There are now a variety of initiatives focused on encouraging people to do more repeated trials of previously published research, especially in the medical, social, and behavioral sciences.

Skepticism: Merton originally wrote this one as "organized skepticism," and he focused more on the habits of skepticism that put scientists in conflict with governments that wanted unquestioning loyalty back in the 1930s and 1940s. Subsequent writers such as Ziman have focused more on how these habits affect the conduct of science itself, and that is probably more what we are interested in here.

How is skepticism in science "organized?" The most institution-alized example is peer review, which we will discuss in more depth later in this book. The basic idea of peer review is that, when you try to publish a paper in the scientific literature, it is first read and crit-icized by other scientists in the same field of research. Peer review is organized and institutionalized criticism, and authors of papers expect tough questions and hard criticism of their work. Nothing is accepted without solid evidence or reasoning, and alternative inter-pretations of experimental results are often suggested. A paper will only be published after the peer reviewers are satisfied that it makes no unwarranted claims or interpretations.

The skepticism that is organized in peer review, however, also shows up in a less organized form in the way scientists talk among themselves. Individual scientists also internalize this skepticism so that, in conversations with themselves in their own heads, there is a lot of questioning of thoughts and conclusions. Once again, honesty with oneself is one of the important forms of honesty that the scien-tist must practice; internalized skepticism is one important way in which this sort of honesty is achieved. It is this internalized skepti-cism that gets you to the sort of "leaning over backwards" honesty described by Feynman.

Communalism, universalism, disinterestedness, originality, and skepticism: these are what some metascientists think scientists value most. Do most working scientists actually know these words and use them? No. Most could not tell you what is meant by the "sci-entific ethos." Scientists were doing science long before sociologists like Merton decided to formulate the ethos; the norms of the ethos are mostly passed down from mentors to their students by example, anecdote, and informal instruction about how to conduct oneself.

Is the idea of the scientific ethos then of any value to a scientist or a science student? I would argue that a simple rule like "Be as honest as you can be with yourself and others" might be a handier ethical guide on a day-to-day basis. But, as we saw in chapter 1, the scientific ethos can help us to understand how science has man-aged to come up with reliable conclusions about the world despite all the philosophical difficulties that we have in defining a single universal "scientific method." It is also helpful to understand how the community of scientists may be different in their values than other segments of the societies in which we live. And it may help us to better explain to nonscientists why we behave the way we do, especially when, as in Nazi Germany or Soviet Russia, the scientists are acting differently from the way political leaders want them to.

2.14 THE CONTEXT OF SCIENCE

It is important to keep in mind that all the rules and values of science formulated by philosophers, scientists, and sociologists are in some sense ideals; in practice, of course, individual scientists, and sometimes science as a whole, may fall short of these ideals. Back in section 2.9, on ethics in practice, we looked at one example of why someone might fail at following the rule of honesty. We had a pharmaceutical scientist thinking about faking data in order to preserve the outlook for a prospective drug, hoping also to preserve their own prospects within the company. What that example illustrated is that science is not done in isolation from the rest of the world. Science is done by real humans, who face all sorts of political, economic, and social pressures. It is therefore worth looking more into the context in which science is done—especially since this context is often what determines what sort of science gets done. Scientists sometimes answer the question "What should I do?" with another question: "Well, what do you *want* me to do?"

In today's world there are three major realms that determine what science is done, by whom, and for what purpose. These realms are academia (colleges and universities), government, and industry. The roles of government and industry in science may seem more prominent than they were 150 years ago, but the realms of government and industry were never completely separate from science.[11] Private foundations and nongovernmental organizations could be considered a fourth realm that both employs scientists and funds science. Although these organizations often disagree with particular governments on some issues, their overall objectives for the most part align with those of government, so I won't be describing them as having a separate set of objectives in the following discussion.

Each of these three realms has its own objectives and its own reasons for supporting, or doing, science. The principal objective of academia is to discover the truth about the world and to disseminate it. The principal objective of government, private foundations, and nongovernmental organizations is to further the health and prosperity of a region, nation, or state and its inhabitants. The principal objective of industry is to make a profit.

These different objectives affect the sort of science the different realms do, or support. For example, a scientist in academia may be interested in the fundamental question of what causes type 2 diabetes. A scientist in government may be interested in how this information can be used to make a healthier society whose members suffer less from type 2 diabetes. A scientist in industry may be interested in developing drugs that could treat type 2 diabetes, especially if (and usually only if) there is a profit to be made from such a drug. All of these scientists are likely to read the same journals and attend the same scientific meetings, and will have a common interest in finding out the truth about the disease. But their behavior—in research, in publishing, and in seeking support for their research—will be somewhat different.

2.15 RESOURCES FOR SCIENTIFIC RESEARCH

It would be simple if these three realms went about their own business, pursuing their own aims, and meeting occasionally at a scientific meeting. While that may have been the norm in the nineteenth century, it is certainly not the case now. Modern scientific research costs a lot of money, both for equipment and supplies and for the labor of the people working in laboratories and in front of computers. The realms with the most money—industry and government—are not necessarily the realms with the people most interested in advancing fundamental understandings of nature. A lot of the interaction between the realms, then, has to do with the transfer of resources—of money—among the different realms.

Many people don't realize just how involved governments are in funding scientific research. In the United States, the National Science

Foundation funds basic science research, the National Institutes of Health funds medical research, the Department of Energy funds energy-related research, the Environmental Protection Agency funds environmental research, the Department of Defense funds military research, and the Department of Agriculture funds agricultural research. Other parts of the government funding research are the National Oceanic and Atmospheric Administration, the National Aeronautics and Space Administration, the Department of Homeland Security, the National Institute of Standards and Technology, the Patent and Trademark Office, and the United States Geological Survey.

What right does the government have to spend taxpayers' money on this research? In a democracy, the government should reflect the will of society as a whole. In most industrially developed nations, people mostly recognize the benefits of science, or at least the benefits of some of the technologies that follow scientific advances, and support the idea that at least some of this scientific research should be funded by the government. As a result, elected representatives of the people then designate parts of the national budget toward funding scientific and technological research. This doesn't mean everyone in a society is happy with the result, or even that science budgets are always allocated based on high-minded discussions of the benefits of fundamental science. The popularity of government funding of science rises and falls, and so do the budgets. But since the middle of the twentieth century, government support of science has been continuous, and it is hard to imagine industrially developed nations ever completely cutting it off.

Although government funding of science follows generally from the will of society to support science in general, the particular research funded by government is not designated by the citizens, and is rarely specified by their elected representatives. Legislatures may specify how much money will go to medical research, energy research, and environmental research, but within each of these categories, scientists make a lot of the decisions about what individual research projects are to be funded. Many of the scientists who decide what research is funded are not employed by the government—they are employed by academia or industry and just serve on committees set up by the funding agencies to recommend whom to fund. Even those employed by the government will have been trained in academia and may have rotated into the government job after some time in academia or industry.

The general population, and their elected representatives, may not know too much about the science being discussed in the grant proposal review committees, or see the connection between rather abstract-sounding research and practical benefits that may ultimately result. As a result, there is always something of a struggle between scientists who want more resources to pursue fundamental questions and people and their governments who would rather spend money on other things—unless they can be convinced that there is a definite payoff to the scientific research in the not-too-distant future. Few scientists go through their careers without some awareness of this continual struggle.

A lot of government money goes to academia, and a large majority of the research that goes on in academia is funded by the government. But government also gives money directly to industry. In the United States, some of this money is in the form of Small Business Innovation Research (SBIR) and Small Business Technology Transfer (STTR) grants, which are handed out by the same agencies and departments that fund academic research. Why does the government do this? Well, sometimes what society needs are innovations that are a bit too applied to attract the interest of academic scientists but a bit too long-term to attract the interest of industry, which needs to worry about profits in the next few quarters. Government, as the realm most committed to providing for society's needs, steps in and pays businesses to do the research and development needed to get new technologies off to a running start.

Another way in which government supports industry in the United States is through the Orphan Drug Act (ODA); drugs that might help patients with rare diseases often aren't actively researched by pharmaceutical companies because they don't represent a lot of future profit. The ODA can give tax incentives, enhanced patent protection, and clinical research subsidies to a pharmaceutical company that is willing to develop drugs for these rare conditions.

Government also funds industry research in other fields, especially defense research and energy research. The Department of

Defense may sometimes prefer industry over academia for this research because the need for secrecy is often incompatible with academia's more fervent adherence to the values of openness and communalism.

There are also relationships between industry and academia. The development of a drug like Alimta, discovered at Princeton University but developed into a drug at Eli Lilly, is an example of such a cooperative relationship. There are lots of different forms these relationships can take, from simple consulting arrangements where a professor is paid a certain amount per hour to advise a company, to much more complex ones where research responsibilities, costs, and benefits are shared according to negotiated arrangements. The ethical issues involved in such relationships can be tricky. Academic scientists may find themselves pressured to come up with results that support the conclusions most beneficial to the industrial partner's bottom line. Also, the industrial and academic partners may have very different ideas of what the norm of communalism requires with respect to disclosure of the results of research.

There are also plenty of situations in which all three realms are closely interlinked. Take, for example, the experimental tokamak fusion reactor at General Atomics in San Diego. It's largely funded by government, it's run by a corporation, and a lot of experiments done there are overseen by professors from the nearby University of California San Diego. In situations like this it can be hard to even discern the dividing lines between the three realms as people work together to solve problems that are fundamental in nature, are likely to benefit society as a whole, and may enable someone to make a profit sometime in the future.

2.16 ETHICAL CONFLICTS

Not all of the interactions among these three realms involve a mutually beneficial flow of resources and innovations, however. There are also a lot of ethical issues that arise out of the different objectives in the different realms. These can create strong incentives for violating the *norms* of the scientific ethos. The academic scientist may want to find out about an interesting property of an organ or tissue affected by type 2 diabetes, and this property may not be directly related to finding a treatment or cure for diabetes. But if the scientist wants to get money from the government, it may be necessary to argue that information about this property is important for understanding the disease and improving public health. You can imagine how, in some cases, the *honesty* of the argument may not be as stringent as what we might imagine scientific honesty to be under ideal circumstances.

Likewise, an industry scientist may not be completely open in publishing preliminary research into the possible mechanisms for a potential drug that alleviates symptoms of type 2 diabetes, because it might give a competitor ideas about developing a competing drug. But an academic scientist's search for answers about the disease may be hindered by this lack of *communalism*; the secrecy may result in more repetitious research that slows the development of scientific understandings.

Finally, government may be influenced by industry lobbyists not to fund some academic research if that research might lead people to cure themselves of the diabetes without the expensive drugs created by industry—another, more subtle, violation of the norms of *communalism* and *disinterestedness*.

So, while the interactions among the three realms provide a flow of funds and ideas that is often productive, this flow can also create conflicts of interest and potential ethical problems. Some scientists will, in their careers, have to confront fascinating ethical issues around human gene editing or animal experimentation. But almost *every* scientist has to confront the ethical issues connected with money or other resources, and for most scientists, these will be the most persistent and nagging ethical issues.

There's no way I can cover all the interactions among government, industry, and academia in a chapter like this; my interest is mostly in giving you a framework that can help you understand the tensions and synergies that can arise when science and society intersect. And it is also important to realize that when we think about "What should we do?" a lot of the time there are people trying to tell us what to do!

FOR FURTHER STUDY AND DISCUSSION

1. Reflect on some decisions you have made in the last year. See if you can come up with examples in which your decisions were based primarily on each of the five types of moral reasoning: consequentialism, social contractarianism, deontology, virtue, and care.
2. In the subsection on consequentialism, we discussed one way of thinking about medical experimentation. But in most countries, the ethics of clinical trials are usually discussed in terms of a social contract that outlines rights for patients enrolled in these trials. Does the existence of this social contract make consequentialist considerations about clinical trials irrelevant? How might deontological, virtue-based, or care-based ethical systems contribute to the discussion about medical experimentation?
3. Sometimes, in discussing ethics, it is tempting to focus on particularly difficult ethical choices. Many ethical choices we face more frequently are far less difficult, but the cumulative result of all these little choices may be as important as one big choice. Try to imagine three ethical choices that you might have to make in your future career as a scientist or professional. Describe one of each of the following:
    a. An example of an ethical choice you might have to make frequently, or are very likely to have to make, but one in which the right thing to do should be fairly obvious.
    b. An example of an ethical choice which you would rarely have to make, or which you might never have to make, but one in which it would be really difficult to determine the right thing to do.
    c. An example of an ethical choice somewhere in the middle; maybe a decision of this difficulty might be made once a year, and might be only somewhat difficult to make.
4. Search the World Wide Web for government granting agencies and find a "Call for Proposals" or "Funding Opportunities." Figure out what kind of research they are asking for, and determine how (or if) it fits into the scheme of the three realms outlined in this chapter. What kind of research are they looking for, and why? Who would most likely do the research?

ADDITIONAL READING

Held, V. *The Ethics of Care. Personal, Political, and Global* (Oxford University Press, 2007).

This is a good overview of the ethics of care, summarizing a lot of previous work and controversy.

Kovac, J. *The Ethical Chemist. Professionalism and Ethics in Science* (Pearson Prentice Hall, 2004).

I am indebted to Kovac for giving me the idea to include ethics in my course in the first place, and for the general ideas of presenting both general ethical theories and professional norms. This book also has a wide range of case studies, from small everyday decisions to major professional issues.

Ziman, J. M. *Real Science. What It Is, and What It Means* (Cambridge University Press, 2000).

Ziman's book is probably the most thorough recent discussion of Merton's norms.

PART TWO

STANDING ON THE SHOULDERS OF GIANTS

CHAPTER THREE

THE SCIENTIFIC LITERATURE: AN OVERVIEW OF THE TERRAIN, AND A BRIEF HIKE IN

LEARNING OBJECTIVES

After reading this chapter, you will be able to:

- correctly identify examples of primary, secondary, and tertiary literature, and describe the relationship between them
- give examples of different kinds of books, and identify which books or other sources will best provide you with the basic understandings critical for further research on a topic
- explain why you need to get a sense of the vocabulary and basic concepts of a research field before getting into the more advanced literature
- explain how library cataloging systems can be used to find books relevant to a research area

3.1 HISTORY, METAPHORS, AND LITERATURE

Any study of the history of science makes it evident that much scientific progress depends on scientists being aware, through manuscripts or published writings, of the successes and failures of past theories and experiments. Isaac Newton is often quoted as having written, "If I have seen further, it is by standing on the shoulders of giants."[1] Newton made his discoveries in the mid-to-late 1600s, the very time at which the development of modern science began to accelerate. People of the time were realizing, thanks to Francis Bacon, that they had to expand their observations of and experiments on the world in order to move beyond the understandings of the medieval Scholastics. But even with this newfound fervor for experiment over scholarship, they also realized that they were building on the discoveries of Ancient Greece, the work of Arab scholars, and the incremental achievements of medieval Europe.

In the twenty-first century, standing on the shoulders of giants is a much more athletic and acrobatic maneuver than it was for Newton. There are generations and generations of giants, all standing on top of each other's shoulders. How do we get up there? The shoulders-of-giants metaphor tells us something about the importance of the achievements of our predecessors, but it tells us little about how to achieve the great elevations needed to see further. I would like to introduce a new metaphor that focuses on accessing the insights of our predecessors. The insights of our predecessors are contained in the scientific literature, and I like to view this literature as a *landscape*; there are many trails through this landscape, and we have to find the one that most comfortably leads us from the lowest valleys to the tallest peaks.

3.2 SUBRAMANYAM'S CYCLE

I am not the first person to use metaphors to describe the scientific literature. If you talk to science librarians about the scientific literature and how it relates to the process of scientific research, you will hear about a "scientific publication cycle," or maybe a "research cycle." This is largely because Krishna Subramanyam, a professor of library and information science, introduced just such a cycle in the 1970s and 1980s to show how scientific information evolved. The original version of this cycle is shown in figure 3.1; you can find all sorts of simplified and brightly colored versions of this cycle on the World Wide Web, mostly at websites run by academic libraries.

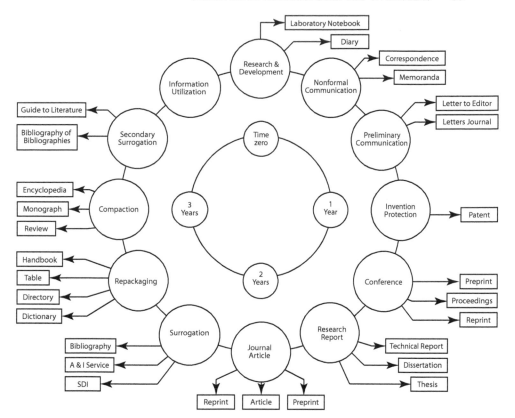

Figure 3.1. Subramanyam's research cycle, as originally portrayed in the 1970s. Copyright 1979 from *Encyclopedia of Library and Information Science: Volume 26 – Role Indicators to St. Anselm-College Library (Rome)* by K. Subramanyam. Reproduced by permission of Taylor and Francis Group, LLC, a division of Informa plc.

This cycle helped science librarians put into perspective all the different resources they were in charge of keeping and helping library patrons to use. It's useful, too, for the science student, who can use it to better understand the process by which all these information resources are generated. I think it is a nice aerial picture of the landscape of the scientific literature—the sort of picture that you would take from an airplane 5,000 meters up in the atmosphere. But if you want to actually learn something from the literature, you can't stay in the airplane at 5,000 meters up. You have to get on the surface of this landscape and wander around.

It's best not to just parachute in from your airplane, either. There are lots of modern search tools that can take you straight to difficult scientific articles in the primary literature. But this would be like parachuting onto a mountaintop without really knowing where you are, how the mountaintop relates to the surrounding landscape, or how to find your way back to where you came from. It's much better to start from a low level and find some gentle valley to climb. Once you have gained some altitude, you can start looking for ways to climb into the higher mountains.

The purpose of this chapter is to discuss how you might find some gentler climbs. Subramanyam's cycle, like an aerial view, can give us hints as to where to go, but I am sorry to report that, for any given area of research, you will need to find your own route—a route that allows you to hike slowly and steadily to higher and higher elevations. You will have to take on the attitude of an explorer, be willing to try different trails, be willing to double back when you run up against obstacles, and be willing to ask for directions from a guide such as a librarian or professor.

But before we talk about your personal journey into this unfamiliar terrain, let's take a look at Subramanyam's map. It has an inner cycle indicating the typical timeline for the incorporation of ideas into different forms of the scientific literature, but I think this varies a lot from discipline to discipline, and even between sub-disciplines. The outer cycle describes what is happening to the ideas and information. The rectangular boxes at the periphery of the map show examples of the written or published output from each of the different steps along the cycle.

Methods, Data, Observations, and Informal Communication: The first steps, starting at the top, involve the research itself. Researchers are writing down methods, data, and observations in their laboratory notebooks. Few scientists work alone, however, and so researchers are also having discussions with their colleagues about what they are finding. Nowadays we would definitely have to include *emails* among the *correspondence* and *memoranda* that we see on Subramanyam's cycle. These communications aren't public; they are only shared within a group of researchers among whom there is a lot of trust.

Primary Literature: The next whole section of the cycle is a lot more complicated than the simple map makes it look. Some research first gets its public exposure at a *conference*, other research may be first published as a *letter* or *communication*, and occasionally some research goes directly to publication in a regular *journal article*. We'll discuss the difference between these different sorts of publications in a later chapter; they all occur in *scientific journals*. *Patents* only result when the research has a potential economic benefit to the researchers, and these are filed with a government agency; *theses* and *dissertations* only occur when the researchers are students, and these tend to be kept on file at universities. The one thing that all of these publications have in common is that they originate with the people actually doing the work, and directly report on the experiments and results of the work, although the level of detail may vary. All of this part of the landscape is called the *primary literature*.

If we really wanted to make an accurate map of this part of the landscape, we'd have to show a lot of different trails. Some research first published as a letter might have a fuller account published later as a journal article—that's how it's *supposed* to work. But sometimes the letter is never followed up with an article if the research doesn't live up to its initial promise. Some parts of a student's dissertation may have already been published as papers; other parts may still be waiting to be published as a future letter. So, it's not a simple cycle with a single path.

Secondary Literature: The next section of the map is about what people do with the primary literature once it's published. People read it, of course, but not everyone can read all the articles that are published in a given area of research. So many *secondary* forms of literature have been invented that try to make it easier for scientists to keep up. Some of the oldest of these are *abstract collections* of all the articles in a given field, such as *Chemical Abstracts* or *Biological Abstracts*. These, along with *bibliographies*, *catalogs*, and *indexes*, are the result of what Subramanyam calls *surrogation*, but what others have just called *organization*.

Compaction is the process that yields *reviews* or *review papers*, *monographs* (books about a particular subject), *textbooks*, and *encyclopedias*. In a review paper or monograph, each paper in the primary literature may only get a paragraph-long summary, but hundreds of recent papers may be covered, giving the reader an overall sense of what has been learned in a field of research. Textbooks and encyclopedias give an even more compressed view of a field; these tend to focus more on older, well-established conclusions in a field rather than the more recent findings included in a review.

Repackaging is what Subramanyam calls the process of just taking particular bits of data from the primary literature and organizing them into *dictionaries, tables,* and *handbooks*. Here we can have handbooks that list the lifetimes of unstable atomic nuclei, tables of the thermodynamic properties of compounds, or taxonomic keys to identifying species. The dictionaries we are talking about here may be defining scientific words, but they may also focus on other data, or even have extended, explanatory definitions. For example, the *Dictionary of Inorganic Compounds* gives both physical information for identifying inorganic compounds and references to articles that tell how to synthesize these compounds.

You may have noticed that I put these in a different order than Subramanyam did; I did so on purpose, as I think that the dictionaries, tables, and handbooks represent the greatest reduction in content from the original literature—in some cases, it may not even be possible to trace what primary literature was used in assembling the tables or handbooks.

It is important to realize that librarians often consider some of these sources to be tertiary sources. I am following Subramanyam's example by calling them all secondary sources, as many examples in each category will directly reference the primary literature. The important thing, however, is not what we call them but how we use them—a question I will address in the next section.

Tertiary Sources: As a chemist, I never had occasion to use the *bibliographies of bibliographies* or *guides to literature* that Subramanyam puts into this category—until I was trying to write this book! One might argue that any resource that directs you to review articles would also be a tertiary source—but most of the time these can be found in the same indexes and abstract collections that are used to find articles in the primary literature. So, unless you are interested in becoming a science librarian, I wouldn't worry too much about these.

3.3 APPROACHING THE LANDSCAPE

Sometimes, when Subramanyam's cycle is presented, the process for using the scientific literature is simply described as a matter of going around the cycle backward. The user starts with the tertiary sources, which would tell the user which secondary sources to use. From the secondary sources, the user can access the primary literature. Subramanyam suggested this general approach, but admitted that the way information is used is frequently a lot more complicated than this. For example, I almost always skip over the tertiary sources. Most scientists gradually learn which secondary sources are most helpful when you are looking for particular kinds of information. Some of that information is right there in the secondary sources, as when you are looking up a thermodynamic value in a handbook or table. Other times, you might use a different secondary source such as an index or abstract collection to find review articles, which themselves are secondary sources. But in the index, you will also find references to primary sources—original research articles—right next to the review articles. You can choose whether to go first to the review articles, which I recommend for the less advanced explorers, or go straight into the primary literature. So, the idea that you just go around the cycle counterclockwise is an oversimplification.

That is why I like to focus on the metaphor of the scientific literature as a hike into unfamiliar terrain. As I said before, you want to find a gentle path. If you're hiking in real mountains, and the path is too steep, your legs will start to ache, and you will begin to feel lightheaded. It's similar with the scientific literature. If you aren't understanding at least half the sentences in the source you're reading, you will get confused and have a headache! If this happens, the trail you chose is too steep. *You have to be able to understand what you're reading.* If you're not understanding, backtrack and see if you can't find another, gentler, pathway into the literature. One common reason for not understanding what you are reading is that there is too much unfamiliar vocabulary. If we are just entering the literature of a particular field of research for the first time, we have to think about what would be most useful to find first. As someone who has explored a lot of literature landscapes, I would say that one of the first things you need to be looking for is *vocabulary*.

Figuring out what words scientists use to communicate in a field of research is important not only for understanding what they write. The vocabulary can also be used to help you search for more relevant information as you go farther into the landscape. For example, let's say you were doing background research for a household science project involving making batteries from coins, paper towels, and saltwater. If you try to get into the primary literature by directly searching using words like "batteries," "coins," and "saltwater," you won't be very successful. But in a textbook section or encyclopedia article on batteries, you might find that, in the context of batteries, the saltwater is serving the role of the *electrolyte*. You may also realize that the coins are really just examples of *metals of different redox potential*, and that they take the roles of *anode* and *cathode*. The battery itself will be an *electrochemical cell*, specifically a *galvanic cell*. These vocabulary terms are going to help you when you start trying to search for review articles and primary literature in the abstract collections and indexes, and even in library catalogs. These vocabulary words might be entirely new to you, or not so new, but even seeing and writing down the more familiar ones should reassure you that you know how the subject of your research is written about in the literature.

Acquiring a new vocabulary isn't just a matter of writing down a list of words; you need to know what the words mean, and in the process of getting that information, you will also get a sense of the relationship between all the different words. If you don't know how the words relate to each other, you need to do a little more preparatory work before you set off in the direction of more advanced literature.

With this in mind, where are good places to find new vocabulary? Because scientific journals are primarily designed to be read by people who are already professional scientists, they are not the best place to start for a student just beginning a research project. Books are a better place to start, but there are many different types of books in the scientific literature, so I will give you a brief guide to them in the next section.

What about online resources? Well, these can be just online versions of journals, or books, or they can be resources only available online. When we look at sources online, it is useful and important to be able to classify them as *electronic books*, *electronic journals*, and *electronic reference works* or *databases*. All the things I say about books below apply to online books as well, and electronic journals will be the equivalent of the journals discussed in the next chapter. Electronic reference works are something different, and in using these, you will have to be careful to evaluate the trustworthiness of the source based largely on who creates and maintains it.

3.4 KINDS OF BOOKS

Textbooks. These are the science books that you are most familiar with, and they might be the best books to start with, because they are often specifically designed for teaching students a new vocabulary. I often find that the "new" vocabulary that students need to understand a project is actually "old" vocabulary that they learned in a course they've already taken—the students may have forgotten the vocabulary or just don't realize that the project they are interested in now is really just an application of what they learned in a previous course. When considering whether to sell your textbooks at the end of a course, keep in mind that these books may well be useful to you in the future, especially since libraries often don't keep standard undergraduate textbooks in their collections!

Monographs. Monographs are the most traditional kind of book; these are books about a single, well-defined subject, usually written by a single author (but not always, especially in the sciences). Monographs are rarer in the sciences than they are in other disciplines. In history or literature, monographs tend to be the standard form of scholarly output, and are seen as being just as important as journal articles, if not more so. In the natural sciences, all of the new knowledge shows up first in journals, and only occasionally do scientists sit back and collect the knowledge from journals into a monograph. More often, recent knowledge is just collected into review papers published in review journals.

Often, a book isn't written in a given field until professors start thinking about teaching courses in the field, so some monographs overlap with the textbook category. These can be intended for graduate-level courses, so they may not be the easiest to understand, but sometimes these are the closest we get to scholarly monographs in the sciences. The prefaces to such books often mention that they have been written to serve dual roles as both a graduate textbook and a resource for scientists new to a field.

Another form of monograph in the sciences is the *popular text*; this is a book written to introduce nonscientists to a discipline or sub-discipline. Popular texts have a long history in the sciences; Mary Somerville, a nineteenth-century Scottish woman, popularized many of the physical, biological, and geographical sciences of her time, was well known by the public, and was well respected by the scientists whose work she popularized. The level at which popular texts are written can vary, but for an undergraduate looking for an easy climb into an unfamiliar territory, they can be very useful.

Conference Proceedings. Some books you see published in the sciences aren't really like the sort of books you encounter in non-academic literature. One such sort of book is a conference proceeding; it collects individual presentations given by different people at a conference into a single volume. Unlike the other books we've discussed so far, which are usually best categorized as secondary literature, a conference proceeding usually should be classified as primary literature, as the authors of the individual presentations are presenting original research. If the scope of the conference is fairly limited, such a book might be considered a monograph, because it's about a single, well-defined subject; on the other hand, you can't expect a unified point of view from the different authors, and the authors may disagree with each other. The quality and usefulness of conference proceedings can vary a lot. At their best, conference proceedings can bring together the best minds in a discipline to give a variety of different perspectives on an important problem. At their worst, conference proceedings can be a collection of unfinished research projects, uneven in quality, that aren't as well peer reviewed as the research that appears in good journals. So, I usually don't recommend conference proceedings as the best place to start a hike into the scientific literature.

Review Collections. Somewhere between the more traditional monographs and the conference proceedings are the collections of review articles that some editors put together on a given topic. Librarians may actually consider these to be monographs, as they are limited in focus to a particular topic, but there's a lot of gray area here. Collections of review articles are typically more helpful than conference proceedings, in that they tend to cover a lot more peer-reviewed research because they are more solidly in the category of secondary literature. They can also be broader than a monograph written by a single person, because no single expert can read, digest, and write intelligently about as much primary literature as a whole collection of experts! Unfortunately, these volumes can be pretty advanced, and are beyond what an undergraduate needs to get a good introduction to a field.

Reference Works. *Reference works* are the sorts of books that are kept in the *reference section* of a library; the reference section is, traditionally, staffed by a *reference librarian* who knows which reference work to go to in order to get a particular kind of information. Reference works or reference books are *non-circulating*; they can't be checked out, so that they're always in the library for people to use. Monographs, which take longer to read, generally don't get put in the reference section, so that they can be checked out for people to read over the course of a few days. What you find in the reference section are things like *encyclopedias, dictionaries, handbooks,* and *tables.* As more and more reference works become available online, however, there is an increasing tendency for libraries to shrink the physical space given to the reference section and to spread the role of the reference librarian over many more library employees. In some libraries, some of the print reference works end up moving to the *stacks* (the place where the books you can check out are kept); in other cases, the print reference works are discarded, especially if a similar electronic resource is easily available.

As a scientist, I tend to view reference works as being of two different kinds: *general* reference works, and *technical* reference works. This is not a distinction that librarians usually make, but it is one that I find very useful. General reference works are designed for a more general audience and contain more general information. Encyclopedias and some dictionaries fall into this category. By "a more general audience," I don't necessarily mean your average person off the street; scientific encyclopedias are typically intended for someone who has had at least a few undergraduate courses in science. By "general information," I mean a broad introduction to a topic, rather than the very particular sorts of data that you would get from a data table or handbook. It is important to realize that the names of reference works are not controlled—there are some things called "handbooks" that are closer to encyclopedias. Likewise with dictionaries—some really focus on particular data, and I would classify them as *technical reference works*, but others have more general information.

Technical reference works are those that are most useful to people who really know what specific information they are looking for. A once-commonly used example of this is the *CRC Handbook of Chemistry and Physics*, which has tables and tables of all sorts of data from the stability of isotopes to the composition of Earth's atmosphere to the melting point of thousands of different organic compounds. It's not the reference book you'd want to look at if you were trying to understand what an *aldehyde* is; it's the book you'd go to if you already understood aldehydes and wanted to look up the boiling point of 4-hydroxy-3-methoxycinnamaldehyde. Technical reference works like these aren't very helpful to students just trying to get their bearings in a new landscape; they're better for finding technical information once you've started climbing some of the higher mountains.

Some of the most useful general reference works are *encyclopedias*. Encyclopedias have a long history, stretching back to the first century CE. Many encyclopedias, from the *Naturalis Historiae* of Pliny the Elder in the first century CE to the *Wikipedia* of the twenty-first century, try to cover the whole of human knowledge. Other encyclopedias focus on particular branches of human knowledge, such as the *Kirk-Othmer Encyclopedia of Chemical Technology* or the *Encyclopedia of Applied Plant Science*. The more specialized encyclopedias may demand a little more background knowledge but usually no more than might be expected of an undergraduate who has taken some science courses, and they may be just what you need to progress further into the landscape.

Of course, the encyclopedia everyone thinks about first these days is *Wikipedia*, the online encyclopedia. *Wikipedia* has really changed the way everyone thinks about information. When *Wikipedia* first started becoming popular, many librarians and scholars were skeptical about using it as a resource, as it seemed too easy to put up unreliable or misleading information. But gradually, starting with a study by the science journal *Nature* in 2005,[2] many people have realized that its accuracy in some fields is at least comparable to that of other secondary sources. There seems to be a growing consensus that *Wikipedia* can serve well for the initial hike into an unfamiliar landscape. In some cases, it is as good as a textbook in providing vocabulary to use in future searches, and in explaining how different parts of a field fit together. Many *Wikipedia* articles have links to other websites that can provide further information, but you have to be careful to evaluate the trustworthiness of these websites. Like more traditional print works, they also have *references to*, or *citations from*, other secondary or primary literature—signposts to other trails leading into the more difficult parts of the scientific literature landscape.

3.5 A PLAN

Given all these different types of books, I would say that the best places for a student to start climbing into the literature are textbooks, some monographs, and some general reference works such as encyclopedias. In heading off to use these resources, however, it is helpful to have something of a plan. If we just start entering words into a search engine, or start thumbing through an encyclopedia, we may be somewhat successful, but a strategy can make our efforts more efficient.

Also, consider the very real possibility that your research will not be done in an afternoon but will stretch over the next few weeks or even months as you learn more about your project and start to embark on the experimental work. You shouldn't just rely on your memory; write stuff down! I require my students to keep a three-ring binder with notes and assignments related to their searches of the literature, just as laboratory scientists keep a written laboratory notebook. With such a binder, you won't forget what you have already learned, and you won't be starting out your climb into the scientific literature landscape at sea level every time you go to the library.

Now, to the plan: first of all, it is useful to pause and consider what you already know. Let's use the example of the proposed household science project on batteries made from coins. Did any of your textbooks cover batteries? Your general chemistry textbook probably did—maybe it would be wise to find that textbook and refresh your memory as to the fundamentals of how electrochemical cells work. Write down some vocabulary words that you may have forgotten but that you think might be useful in further researches. Draw a diagram of what happens in an electrochemical cell to try to picture the relationship of the different vocabulary words.

Once you have a good sense of what you already know, figure out what you need to find out. Perhaps the textbook emphasized the *theory* of batteries far more than the *practical aspects*, and you might need to find a more practical source. An encyclopedia article on batteries might give you more information about how commercial batteries are constructed, and this might give you some answers, and more questions. You may wonder if you would get different results with different coins. It would then be important to find out what metals are present in different coins. There's probably some reference work that can help you figure out what metals different coins are made of, or a reference work that will tell you what other common items contain the particular metals you think you need. With each unanswered question, try to *imagine* where you might find out more information that will help you. I've always believed that imagination is one of the most important skills for a library researcher; you don't tend to go looking for things until you imagine that they're out there somewhere.

As you go through encyclopedias, books, and websites getting answers to your questions, one thing you might want to pay attention to is what kinds of experimental approaches people often use in a given area of research. Common experimental approaches are part of the *research paradigm* that Kuhn described as shaping science in a particular field, as we covered in chapter 1. One thing that might be useful is to keep a list of *observables*, the qualities or parameters that help scientists characterize what is going on. For example, in doing research on batteries, you may find that people are interested in observing current, voltage, or time to failure. These are things you can observe, and even quantify, in your own experiments. If you find examples of experiments very similar to what you are thinking of doing, pay particularly close attention to these, and take good notes. Think about how you can expand your experiments beyond what has already been done by others.

Note that this "plan" isn't really a simple three-step plan—you will have to repeat the steps of reviewing what you know, asking questions, finding vocabulary, and thinking about observables and experiments over and over. Throughout this process, keep writing down as much as you can; new vocabulary words, sketches that help you relate the vocabulary words to each other, new questions and the answers to them, and possible observables and experiments. Also, write down any *references* that are *cited* by the books, or websites, that you read. Some of these may be references to books or general reference works that are suitable for this initial easy hike into the information landscape; others may be journal articles that you can save in your notes for when you're ready to climb more difficult terrain.

3.6 FINDING BOOKS AND REFERENCE WORKS

Implementing the plan I've outlined above requires that you have some way of actually finding reference works and books. You've probably had some previous training and experience in that. But it may be helpful to look a little at the different cataloging systems that there are and how best to use them.

As soon as large collections of literature started showing up in human history, there were efforts to distill and organize information. The library of Alexandria, founded in the third century BCE, had a cataloging department. Library catalogs are a form of *index-ing*, guiding the library patron to find what the library has in its collection. There was a long stretch of time between the efforts at the Alexandrian library and anything that looks much like what we have now, however. During this period, individual libraries would typically have their own systems for keeping track of their collections, but these usually focused more on where the books or scrolls were kept, and when they were acquired, than on what they were about. In the nineteenth century, however, as public libraries became more common and the number of books in libraries increased, a number of standardized systems were devised that began to be adopted by many libraries.

These systems were based on the subject matter of the books, rather than the location of the books. The Dewey Decimal Classification (DDC), devised by Melvil Dewey in 1876, was one such system; the Library of Congress Classification (LCC) system was devised by Herbert Putnam, the eighth librarian of Congress, in 1897. The Dewey system tends to be used in most public libraries, while most academic libraries use the LCC. Librarians will be happy to discuss with you the relative merits of these two systems, but both do an adequate job of classifying books so that individual libraries can catalog the books and other resources in a library's collection. If we want to find books on a given topic, an obvious destination will be the catalog of our local library. With computerization, we now are able to search these catalogs in more different ways than we once did. Back when catalogs were all on cards, we were limited to searching by author, title, or subject headings that were assigned according to the LCC or DDC. Now we can use words in a title, keywords, or even call numbers. Nevertheless, it is worth understanding how the original classification systems were developed.

When the Library of Congress gets a new book from a publisher, they have to decide where to put it in the LCC. What is the *subject* of the book? If everybody who worked for the Library of Congress had their own ideas about what a suitable subject description might be, there could be chaos. So, the classification systems of the late nineteenth and early twentieth centuries developed what are called *controlled vocabularies*; only certain terms were allowed as subjects.

Then the employees of the Library of Congress just had to decide which subjects best fitted the book, or, in some cases, collectively decide that a new subject was necessary.

We can use these classification systems to our advantage. For example, we might search on a keyword that gives us a variety of books, some of which are useful and some of which clearly aren't. If we can go to the detailed entry for one of the books that *is* useful, we can find out what LCC or DDC subject headings were assigned to it. Sometimes, in an online catalog, you need only click on these subjects as active links to take you to other books that are under the same subject heading. These books might not even be in the same part of the library as the first book you identified as useful; for example, books under the subject heading *Electrochemical analysis* include books with LCC classifications starting with QC (Physics), QD (Chemistry), and QP (Physiology). So, the subject headings can be a great way to get more relevant books, even if you didn't pick the best keyword to start your search.

The Internet and the World Wide Web have made it much easier to access books in other libraries through InterLibrary Loan (ILL). This can be through a local network of libraries that share catalogs and access to books, or through a global catalog such as WorldCat. WorldCat doesn't allow you to request books from global libraries, but it does let you know what books are out there. These books can then be requested through your library's ILL.

When you're thinking about requesting a book from another library, it is good to use whatever data is available to imagine just how useful a book might be. Your ability to judge books by their catalog entries will improve with experience; you may have to order a few duds through ILL before you realize just what you can, or can't, get from a certain type of book. Here are a few things to check when you're looking at a detailed catalog record:

1. Monographs will sometimes have a table of contents shown on the detailed record, giving you a sense of the scope and level of detail of the book. Some might show different authors for each chapter, in which case you know it's more like a collection of reviews or unrelated research reports.

Table 3.1. Library of Congress subclasses for a variety of scientific and technical subjects, and their approximate Dewey Decimal equivalents.

Subject	Library of Congress Subclass	Approximate Dewey Decimal Equivalent
General Science	Q	500
Mathematics	QA	510
Astronomy	QB	520
Physics	QC	530
Chemistry	QD	540
Geology	QE	550
Natural History—Biology	QH	560 (Natural History); 570 General Biology
Botany	QK	580
Zoology	QL	590
Human Anatomy	QM	610 (in Medicine)
Physiology	QP	610 (in Medicine)
Microbiology	QR	610 (in Medicine)
Technology (General) (Many other subclasses under T for different types of engineering and technology)	T	600 (includes Medicine); also 700 Arts, which includes some arts (such as Photography) that are included as technologies in the LCC.

2. Conference proceedings will often have the subject heading "Conference proceedings," or "Subject – Congresses."
3. If it's proceedings from a conference, was the conference sponsored by a major scientific society and were the proceedings published by that society or, if there is no indication of a society affiliation, by a commercial publisher? You need to have a fair amount of experience with different publishers and scientific societies before this information can be much help to you.
4. Is it part of a series? Is it a series sponsored by a scientific society or a national institute? Once again, using this information to evaluate books requires some experience with different series or different societies.
5. Who is the publisher? Is it a scientific society, a university press, or a commercial publisher? This information is hard to do much with until you have some experience, and it can never be relied on

too much, but with time you can get a sense of the relative level of quality and importance of work published by different publishers.

Don't be afraid to experiment with books that you're not sure about. It might be a good idea to browse the shelves of your *local* library to get a better sense of what kind of books are out there. Although you can increasingly "browse" libraries electronically using online catalogs, this usually only gives you catalog entries for books next to each other on the shelves. It's not really a very good substitute for picking up a book, opening it, and seeing what is in it! Browsing through the actual physical collection will improve your ability to more accurately imagine what books in distant libraries might be like. Browsing the shelves can also give you a sense of how the classification system (LCC or DDC) treats the sort of subjects you are interested in. Table 3.1 shows some of the codes you'll want to know about.

FOR FURTHER STUDY AND DISCUSSION

1. Describe situations in which Subramanyam's cycle is better for explaining the scientific literature, and other situations in which the author's metaphor of an information landscape is more useful.
2. Why is it important to know about all the different types of books there are?
3. Investigate your library. What cataloging system is used? How big is the reference collection in your discipline? Are there particular reference works that are a good entry point for climbing into the literature landscape in your discipline?

ADDITIONAL READING

Subramanyam, K. *Scientific and Technical Information Resources* (Marcel Dekker, 1981).
    This is a good summary of Subramanyam's overall view of scientific literature; the material on particular sources is out of date, but the overall scheme is still a useful framework.

CHAPTER FOUR

SCIENTIFIC JOURNALS, PAST AND PRESENT

LEARNING OBJECTIVES

After reading this chapter, you will be able to:

- identify how old scientific journals are, and explain the role of scientific journals in the development of science
- explain the similarities and differences between early scientific journals and the modern scientific literature
- identify the different kinds of scientific literature in modern journals, including the wide variety of material found in more magazine-like journals such as *Science* or *Nature*
- identify trends in scientific publishing that may change the nature of scientific journals in the future

4.1 THE HISTORY OF SCIENTIFIC LITERATURE

The history of modern science is intimately connected with the history of the scientific literature. Humans and their ancestors have been experimenting with the manufacture of stone tools, fire, music-making instruments, paper, gunpowder, medicinal plants, and agriculture for a very long time—the first stone tools are more than a million years old. But it has only been in the last half-millennium that there has been a unified, self-consistent body of knowledge and theory that we modern humans readily recognize as science. One major reason that this has only arisen in the last 500 years is that, with the copying of manuscripts and the invention of printing, it became a lot easier to widely communicate the results of experiments or observations and the theoretical thoughts people had about those experiments or observations.

Without the technology for easily communicating results and ideas to people far and wide, what an individual learns about the physical world might only be passed on to people he or she knows personally. That's usually good enough for the preservation and advancement of technologies, such as the manufacture of paper, the smelting of metals, or the construction of large buildings from stone. A master of one of these technologies could pass along his or her knowledge to apprentices, and an apprentice could, after studying with multiple masters, become a master and take on new apprentices. Each master may have made some improvements in the technology, and each technology would thus have been improved. This is probably how technologies such as making tools from stone, making bronze from copper and tin ore, making paper in ancient China, and building stone cathedrals in medieval Europe progressed. Knowledge passed from masters to apprentices was often considered to be secrets shared only with a few people. This helped keep prices high for the services of those people who knew those secrets.

What characterizes modern science, however, is big, general ideas that can be applied to lots of different technologies, and to the natural world, as well. Transmission of technical tricks from master to apprentice isn't conducive to coming up with broadly applicable ideas about the universe. What the blacksmith knows about metals at high temperatures might be combined with what the glassblower knows about glass at high temperatures to gain a broader understanding of matter at high temperatures. But this won't happen if the glassblower and the blacksmith are in different guilds and don't talk to each other. More importantly, if neither talks to the natural philosopher, who may be more comfortable with abstract concepts and mathematics, no unifying, quantitative theory is likely to develop. The natural philosophers who came before the seventeenth century were always enthusiastic about making general theories, but, as Francis Bacon pointed out, they often weren't using enough data from enough observations of the real world. One important contribution to the great advances in science in the 1600s was the improvement in communications between people. Getting the people who worked with real physical problems on a daily basis—the blacksmiths and glassblowers—to share information with the people who wanted big universal understandings was an important part of the development of science.

The precise details of how experimental and technological approaches were brought together with philosophical and mathematical approaches are complex and the subject of much debate between historians of science. But modern science is possible in part because people with different experiments using different technologies frequently communicate with each other and with the theoreticians who concern themselves with ideas and equations. Scientists might accomplish this by having many individual conversations, but that isn't very efficient. If you were an experimentalist, your best bet would be to talk to *all* the theoreticians, because you won't know which theoretician is going to see the importance of your experiment for the development of a new theory. If you are a theoretician, you won't know in advance which experimentalist has the technical knowledge to make the best experimental test of your theory. The best thing you can do is spread your thoughts and results as broadly as possible and hope that the person best able to make use of them will see them. This is what *publication* accomplishes.

4.2 DID MODERN SCIENCE START WITH GUTENBERG?

Some people have suggested that the printing press was an important development that helped catalyze the beginnings of modern science.[1] The development of the heliocentric model of the solar system by Copernicus is seen as a key event in the start of modern science. Nicolaus Copernicus was a Polish student at the University of Bologna around 1500. His teacher, Domenico Maria Navara, had him study a Latin translation of the Egyptian astronomer Ptolemy's *Almagest*. Navara thought Ptolemy was wrong; Copernicus, studying the *Almagest*, came to the same conclusion, and spent thirty years working in his spare time to write his argument for his own new theory. His book, *De Revolutionibus Orbium Coelestium*, was eventually published just weeks before his death. It inspired the work of a lot of other scientists—Tycho Brahe, Johannes Kepler, Galileo Galilei, and Isaac Newton. So much of modern astronomy and physics may have started with Copernicus reading an old Egyptian work.

What does this have to do with Gutenberg? Well, Gutenberg invented the printing press, and in 1454 he published 300 copies of the Christian Bible. In the next few years, many other works that had been preserved in hand-copied manuscripts saw their way into print—including Ptolemy's *Almagest*, first printed in 1462. Had the printing press not been invented, it might have been harder for Copernicus to access Ptolemy's work, and it might have been harder for Copernicus's ideas to spread to other scientists through publication of his own book.

But if you've ever studied history in much depth, you know that history is rarely as simple as the stories we sometimes tell. In fact, a lot more than the invention of the printing press had to happen before Copernicus could think up, and publish, his great new ideas about the Sun and planets. Manuscripts of ancient thinkers such as Aristotle and Ptolemy were being faithfully copied over the centuries, and were translated from Greek into Arabic, from Arabic into Latin, and from Greek into Latin. Without the translations into Latin in the tenth to twelfth centuries CE, many of the ideas of the ancients would never have made it to the students studying in the new universities that arose in medieval Europe. The development of systems for efficiently copying manuscripts by hand, both in and outside the universities, was probably at least as important as the development of the printing press in getting the manuscripts of the ancients into the hands of clever students like Copernicus.[2] But the impact of all these developments—translating, copying, and printing—points strongly to the importance of communication and dissemination for the development of scientific ideas. Without a readable, widely disseminated scientific literature, it is much harder to build on the thoughts and experiments of others—to stand on the shoulders of giants.

4.3 THE RISE OF SCIENTIFIC JOURNALS

The first scientific journals were published in the middle of the seventeenth century. From before Copernicus and through the time of Galileo, ideas were mostly published in books. But even with the printing press, books were still expensive and rare, and they took a long time to produce. People interested in science also needed more informal ways of communicating. In England, around 1645, some wealthy and educated individuals interested in natural philosophy met weekly, in a group called the "Invisible College," to discuss experiments and ideas. In 1662 this group of individuals received a royal charter from King Charles II, and became the Royal Society. By 1665 the Royal Society had also received permission from the king to begin regular printed reports of the doings of the society. The *Philosophical Transactions*, published in London, thus became one of the first scientific journals.

The Royal Society's motto is *Nullius in verba*, roughly meaning "don't take anyone's word for it." This reflected the spirit of Francis Bacon, who emphasized rejection of received wisdom and convention in favor of more direct observation and experiment. But the fact that the *Philosophical Transactions* got its start so soon after the founding of the Royal Society suggests that even the members of the Royal Society realized that they were willing to take each other's word for some things, or else there'd be no sense in publication. If we are to stand on the shoulders of giants, we are going to have to take the giants' word for it to some extent. (See chapter 2 for the importance of honesty in making the literature trustworthy!)

Reading through the first few years of the *Philosophical Transactions* is a fun and eye-opening thing to do. You can access the first 200 years' worth online at the Royal Society website. It is interesting to see that some of the contributions to this journal are in the form of "letters." This form has been preserved in the scientific literature for more than 350 years; you can still find "letters" in many scientific journals, although they don't tend to have salutations and closings like the letters in the original *Transactions*. But many of the other contributions to the *Philosophical Transactions* don't look much like what we are used to seeing in modern scientific journals.

Henry Oldenburg, who put together the first volumes of the *Philosophical Transactions*, had no prior *scientific* journals to model the *Transactions* on; his models were more likely the journals, newsletters, and pamphlets about all sorts of political and philosophical topics that were circulating in London at the time, especially in the popular new coffeehouses. The journal was an ideal form for getting the thoughts and results presented by oral communication at the Royal Society's meetings to a wider audience, and it was faster than the publication of books. Science and journals turned out to be well-matched; there are now thousands of scientific journals, and in most disciplines, journals are much more important than books.

Oldenburg and other members of the Royal Society could have, in their written *Transactions*, adopted the styles and conventions of academic writing that had been developed in the previous centuries, but they did not. Much of the academic discourse in late-medieval Europe was exactly the sort of thing that Bacon had criticized as fruitless and corrupted; what seems to have replaced it is a very practical and matter-of-fact style, except for the exceedingly polite and flowery introductions to many of the contributions. There are very few citations of previous work in the *Transactions*, in contrast to the many citations found in medieval literature. In this way, Oldenburg's new journal was far closer to other newsletters and journals that were published and circulated in the coffeehouses of London.

Some of the contributions to the *Philosophical Transactions* in these early years are nothing more than brief observations, such as the observation of haloes about the moon.[3] Others are more lengthy "accompts" (accounts) of observations or practical knowledge, such as an anonymous account of tin ore mining and processing in Cornwall and Devon.[4] This latter contribution does exactly what I wrote before *wasn't* done much prior to the development of modern science; it relates to the natural philosophers the very practical

knowledge (in this case, mining and metallurgy) that had previously only been known to practical people working in a trade but that
had the potential to inform theories of geology or chemistry. It may
be that the anonymity of this particular contribution was due in part
to the fact that some of this knowledge was considered trade secrets,
the publication of which would get the author in trouble.

The *Philosophical Transactions* was not the only scientific journal
founded in 1665. Across the Channel, the French *Journal des Sçavans*
also made its debut. Several scholars who have studied a random
sampling of seventeenth-century articles in both journals find
that the authors of articles in the French journal were more likely
to apply mathematics to their findings, or to explain observations
with a mechanism or theory. These differences were fairly minor,
however, and the authors of the study suggest that all the articles in
the random sample probably could have been published in either
journal, given adequate translation and minor revision.[5]

4.4 THE EVOLUTION OF THE SCIENTIFIC JOURNAL AND THE SCIENTIFIC ARTICLE—THE EIGHTEENTH TO THE TWENTIETH CENTURIES

Over the centuries, scientific articles, and the journals that published them, gradually came to look more and more like the articles
and journals we see today. Articles became more concerned with
measurement and theory, and fewer of them just reported observations. As the arguments in scientific articles became more complex,
articles were more and more likely to have more formal introductions and conclusions, and headings within the articles helped to
guide the reader through increasingly complex sets of observations, experiments, and explanations. The use of graphics became
more common, particularly the use of the Cartesian graph that is a
mainstay of modern science (we'll cover Cartesian graphs more in
section 10.3). Titles became more specific, as researchers focused on
increasingly narrow topics. The writing style became less flowery
and personal, and personal details of who observed what, where,
were replaced with details about instrumental apparatus and

experimental methods. Discussions of what experimental results or observations meant for theory were increasingly written in cautious, hedging language—"these results would seem to suggest"—rather than as simple statements of fact.

As science came to be done by more and more people, there came to be a greater and greater diversity in scientific journals. The oldest and most general journals of learned societies (such as the *Philosophical Transactions* of the Royal Society of London or the *Mémoires* of the Paris Academy of Sciences) continued to publish science in increasingly lengthy, detailed papers. But these journals were both slow to publish and hard to get published in; in the nineteenth century, there was a blossoming of more specialized journals that were easier to get into and that published new discoveries more quickly. For example, chemists in London or Paris would increasingly publish in the chemistry journals published by national chemical societies, even if they were members of the more prestigious general academies or societies. The Royal Society and the Paris Academy responded to this competition by launching their own more frequent publications such as the *Proceedings of the Royal Society* or the *Comptes rendus* of the Paris Academy, where the contributions were shorter. Other nations and regions in Europe began their own publications, both general and discipline-specific.

The tension between giving a full and thoughtful accounting of results, observations, and interpretations and the limitations of space and time that confront the publishers is a theme that reappears throughout the history of scientific publication. Many scientists think of this as being primarily a problem that arose in the twentieth century, but it already was something of a problem in the early nineteenth century. By this time, more and more people throughout Europe were *professional* scientists rather than simply well-off nobility who pursued science as a hobby. The first modern research universities were founded in Germany in the nineteenth century, and in these universities research results, and publications, were valued as much as teaching. This created a whole new class of people who were anxious to publish. The research university model quickly spread to other countries and was dominant by the twentieth century, leading to more and more scientists with work to publish. An even greater jump in the amount of science to be published occurred after World War II; the might of the atom bomb and the number of lives saved by penicillin convinced governments worldwide that science was worth supporting with money. All the scientists who got government money to do science had to publish their results, and the scientific literature exploded in size.

With the explosion in the amount of science being published, it was no longer possible to carefully explain and depict every experiment, and to publish every data point. Nor was it conceivable that anyone wanting to keep up with an area of research could read fifty-page articles from everyone working in the field. Scientific articles needed to be standardized and shortened. Throughout the rest of the book we will see the huge effects this has had on how scientific literature is written, read, and understood.

It was around 1940 that the standardized "IMRAD" paper format started to be used. IMRAD stands for *Introduction, Methods, Results [A]nd Discussion*. This standard format for articles allowed a reader who had too many articles to look at the option of going straight for the parts of the paper he or she was most interested in. The drive for standardization was probably most prominent in the biomedical sciences, but other disciplines have adapted and adopted similar schemes. The American Chemical Society, for example, frequently recommends using *Introduction, Experimental Section, Results*, and *Discussion* in papers published in their journals. There is also a lot of variation in whether the *Results* and *Discussion* are kept as two separate sections or combined.

One very important part of a scientific paper that isn't included in the IMRAD initialism is the abstract. This is a fairly recent addition to the scientific paper that got its start in the first half of the twentieth century and that is now standard and required almost everywhere. This addition was also developed in response to the huge flood of articles being published. It allows a reader to very quickly grasp at least what was done (essentially a short Methods section) and what was observed (a summary of the Results); often it also gives some introductory background and discussion, as well. We'll go more into the history of the abstract when we discuss abstracting and indexing in chapter 5.

4.5 WHAT CAN BE FOUND IN SCIENTIFIC JOURNALS TODAY?

It is important to note that the drive for standardization and brevity never completely took over the scientific literature. It was widely adopted for standard *articles*, but even in articles there is a lot of variation from discipline to discipline and journal to journal. In the more magazine-like journals like *Science* and *Nature* the standard IMRAD format is not used to organize the articles, even though most of the same information is still present. In more traditional journals the format might be more strictly enforced. Regardless of the format, an article in general will be a complete treatment of a series of experiments helping to establish some sort of argument. We'll discuss arguments more in chapter 7.

But not all scientific results are reported in articles. *Letters* or *communications*, which were common in the first scientific journals, continue to be an important part of scientific communication. Letters aren't expected to be as complete as articles, and are usually published more rapidly, especially when exciting results are to be communicated as quickly as possible. There are whole journals devoted solely to publishing letters and communications. Many journals specify that letters are expected to be followed up by articles that explain the methods in more detail and have a fuller discussion of what the results mean, but there's no mechanism to enforce this expectation. The result is that there are a lot of letters published with interesting results, but the follow-up article with detailed *Methods* and *Discussion* never appears. This can be especially frustrating if one is trying to reproduce the results reported in a letter or communication!

Another form of scientific communication found in scientific journals is the *review article*. Review articles first appeared in the second half of the nineteenth century but became quite common in the twentieth as the scientific literature exploded in size. In fields that are very active, hundreds of articles and communications might be published that are all relevant to an important scientific question. Not every scientist working in the field can read every one of the papers published—if they did, they might not have time to do much original research of their own. So, a few scientists who know the field very well—or who *want* to know the field very well—might sit down and read hundreds of papers in the same field and summarize their most important conclusions. A good review paper may even comment critically on what has been read, emphasizing research papers that are thought more likely to have solid conclusions, and cautioning against trusting the results of some papers that appear to have methodological or interpretive flaws. A review paper presents no new research and no new experiments.

It is important to know if the article you're reading is a review article; if you want to give proper credit for prior experiments or observations, make sure you're citing the people who originally did the work, and not just the author of a review paper in which you first read about those experiments! There are several ways to know you are reading a *review article*: (1) Some journals will label the article as a review article. This is especially true in publications like *Science* or *Nature*, where most of the articles *aren't* review articles. (2) Some journals only print review articles; these journals will have names like **Reviews** *in Biotechnology*, **Progress** *in Inorganic Chemistry*, or **Trends** *in the Biomedical Sciences*. (3) A review article will have no *Methods* or *Experimental Section*, and will have many—often hundreds of—references.

The three publication types I have reviewed so far—*articles*, *letters* or *communications*, and *review articles*—form the bulk of the scientific literature found in most scientific journals. But other things can be found in scientific journals, and there are some journals—which I will call magazine/journal hybrids—that contain many more different things, such as news items and opinion pieces. *Science*, a publication of the American Association for the Advancement of Science, and *Nature*, an independent publication headquartered in London, are the two most prominent examples of this magazine/journal hybrid.

One thing that has been around since the first year of the *Philosophical Transactions* is the *book review*. Despite the dominance of journals over books in many scientific disciplines, books are still published in the sciences. Some of these are textbooks, and some of them are *monographs* in which one author reviews a field.

Book reviews of these sorts of books can appear in more traditional scientific journals like the *Journal of the American Chemical Society*, but they are regular features of the magazine/journal hybrids, and each issue of those periodicals will usually have at least one or two.

One thing that got started in the magazine/journals, but that is being adopted by some more traditional journals such as the *Proceedings of the National Academy of Sciences (USA)*, is what I call the "non-specialist introduction." In a magazine/journal like *Science* or *Nature*, articles and communications can come from any field. Psychology papers might be found right next to physics papers. The result is that many people who subscribe to *Science* or *Nature* can't understand many of the articles printed in the journal, because they are in a field very different from their own. Rather than have the authors of the articles "dumb down" their articles for the general reader—which would frustrate the serious readers in that field of research—the journals invite other scientists in the field to comment on the significance of the work and explain as much of it as they can to the more general science reader in a non-specialist introduction. It's definitely worth it for science students to know about these non-specialist introductions, as they provide an entry point into interesting science that's more accessible for the beginner than the actual articles. The most difficult aspect of these non-specialist introductions is finding them, as each journal has its own name for them. *Science* now calls them "Perspectives"; *Nature* calls them "News and Views," which is especially confusing, as there are other things in *Nature* that are really more news items, found under the heading "News in Focus." And there's no guarantee that they'll be called the same thing in the future.

News articles can also be found in publications that are published for scientists. Scientists want to know what is going on with government funding of science, or what new telescopes are being built; they also want more scientific views of the events that are reported in other media. *Science* and *Nature*, the magazine/journal hybrids, usually put these news articles ahead of the more traditional scientific articles or letters in any given issue. News isn't only found in these hybrids, though; many scientific societies have stand-alone periodicals devoted mostly to news. The American

Chemical Society, for example, has *C&E News*, originally published as *Chemical and Engineering News*. The American Physical Society has both *APS News* and *Physics Today*. Like the general magazine/journal hybrids, these discipline-specific publications present news items, opinion pieces, book reviews, and blurbs about what exciting science is being published in the more traditional journals. These aren't the sorts of publications you are likely to cite in a scientific paper (although they are sometimes cited), but they are a good way of learning about what other scientists are doing in the three realms of academia, government, and industry (see chapter 2).

4.6 WHAT ABOUT THE FUTURE?

The World Wide Web (WWW, or "the Web") has changed the way scientists communicate, and will continue to do so. One of the most obvious changes is that almost all scientific journals can now be accessed via the Web. Another important change that is less obvious to the casual reader of scientific journals is that, although journals still emphasize the need for concise and standardized communication, there is now the opportunity to provide more data and methodological details in *online supplementary material*. This material doesn't take up precious space in a print journal on library shelves, and electronic storage and distribution of such data are relatively inexpensive. Video data and animated simulations can be shown.

The possibility of video has also created a new type of journal that exists only in video form. *JoVE* is a video journal that focuses on videos showing experiments, and was founded by a researcher frustrated by the difficulty of understanding how to do complex and tricky experiments from the brief Methods sections of published papers.

Peer-review practices of some journals have changed to take advantage of the openness of the Web. The Web makes it possible for more scientists to publicize their views and critiques of papers openly, rather than having a few reviewers and an editor decide what gets widely known. We'll look at this more in chapter 8.

For now, journals are still the most important means of communicating scientific results and ideas, whether they are printed on paper or distributed on the Web. But will this always be the case? Will blogs or something like them become an important way of sharing ideas? Will there be other forms of more rapid communication than the journals? It is always hard to know what the future will be. If Henry Oldenburg, publisher of the first issues of the *Philosophical Transactions*, had had a chance to see what scientific journals would become, he would have been surprised. I think we would be surprised, too, if we could see what journals—if they still exist—will be like 100 years from now.

4.7 CLIMBING INTO THE JOURNAL LITERATURE

There are about as many ways to do this as there are to investigate a mountain range. Yes, I'm getting back to our metaphor from chapter 3, in which I likened the scientific literature in a given field to a landscape with valleys and mountains. We decided in that chapter that the key thing was to find a relatively easy way from sea level to the mountain peaks. In that chapter we also found that *some* secondary sources were good places to start, and that some of these secondary sources had references to other secondary sources or even to the primary literature. The next chapters will give you some more tools for finding journal articles. The following are some ways to go further into the landscape:

General Library Databases. It is becoming increasingly common that library Web pages will not only have ways of searching for books with keywords, but also will have ways of searching for articles with keywords. These are not very powerful tools, but they might get you something to start with.

The WWW. A general search engine might get you something just by entering the right combination of search terms, but that's something of a long shot. A database confined to scholarly works, such as *Google Scholar*, is likely to be more productive.

Publisher Databases. Some publishers have tools for searching all the journals they publish; for big scientific publishers like Elsevier or

Wiley, these can yield a lot of results on a lot of diverse topics. For more focused publishers, like the American Chemical Society, you get exactly what you'd expect—articles that are more related to chemistry.

General Science or Disciplinary Databases. Some of these are free to users; others require subscriptions. One of the most popular public databases is *PubMed*, administered by the National Library of Medicine; as you might expect, it leans toward biomedical topics. But you may be surprised by just how broad its coverage is. Individual disciplines such as chemistry or biology often have their own databases, such as *Chemical Abstracts* or *Biological Abstracts*, which we will discuss in chapter 5. In chapter 6, we will also learn about a very general resource, *Web of Science*, which started primarily as a way to find papers that referenced a given paper. This database also does general searches, although probably not as powerfully as those that are adapted to the special needs of individual disciplines.

Expert Recommendations. This is often how students start their research with a faculty advisor. Graduate students are given journal articles by a variety of professors even before they choose which professor will be their advisor, so they have some papers to work with from the very beginning of their research.

4.8 WHAT'S IN A NAME?

If you are trying to find an article using some of the databases listed above, chances are you will be finding more articles than you know what to do with. One way of being selective about which articles to spend your time looking for, and looking at, is to pay attention to the journals that they are in. Some journals have more important research than others. One way to find out which journals have the most important information is to look up the *impact factor* for the journals. The impact factor is a measure of how often articles in that journal are cited by other articles in the same journal and other journals. These impact factors are compiled by the *Institute for Scientific Information*, which we'll learn more about in the next chapter. The more often articles are cited, the more likely they are to be important. But the impact factor is probably both too precise and too inaccurate a measure of what articles you should be looking at. Too precise because it's given as a very precise number, but too inaccurate because it's really not telling you which articles are going to be good introductions to a subject.

Instead, I recommend the following rough guideline: the more general the title of the journal, the more important its research is, and the better it will be as a resource for finding more information. For example, *Science* and *Nature* publish articles that they think will be of interest to scientists in all sorts of fields; these will usually be very important articles in their own disciplines, as well. What's more, because the authors and editors of *Science* and *Nature* articles know that the articles will be read by scientists from many fields, they will often be published with a *non-specialist introduction* (see above) and will have some general papers cited in the first paragraph or abstract of the article. Following the cited references from these journals (see chapter 6) will help you find more detailed information about methods and prior research in the field.

One level below the journals that cover all sciences are the ones that encompass a whole discipline. The *Journal of the American Chemical Society (JACS)*, for example, covers all of chemistry: physical chemistry, biochemistry, inorganic chemistry, organic chemistry, analytical chemistry, and just about anything else that could be classified as chemistry. Other chemistry-wide journals are published by other publishers in other countries, but in general the articles accepted by these journals are ones that would be of interest to a wide variety of chemists.

Another level down are the journals that are more specialized; the American Chemical Society, which publishes *JACS*, also publishes *Inorganic Chemistry*, the *Journal of Organic Chemistry*, the *Journal of Physical Chemistry*, *Biochemistry*, and others. Articles published in these journals are important to people working in those disciplines, but they aren't important enough to capture the attention of chemists in general.

There are levels below this. Some journals can be extremely specific, such as *Solid State Ionics* or the *Journal of Mass Spectroscopy*. The papers in these sorts of journals will be mostly interesting to a relatively small number of researchers. But you should not assume that they are not at all important. Science progresses by the accumulation of lots of information, and sometimes these very specialized journals are where the details of methods and experiments can be found. In my own research, I have frequently found that fully understanding a paper in *Science* may require reading more detailed accounts of methods and previous experiments in the *Journal of Crystal Growth*. We'll learn more about tracking down this information in the next two chapters.

This general strategy for estimating the importance of journals is not foolproof. Sometimes very important research gets published in rather narrowly focused journals, just because reviewers and editors at the more competitive journals failed to appreciate its value. Also, with the ease of publishing on the World Wide Web, another class of journals has arisen that often have general titles but are of little value. These are *predatory journals*, which are really more money-making schemes than scholarly journals. As with many reputable scientific journals, authors are charged a fee for publication in the journal, to cover production costs not adequately covered by subscriptions. Unlike their more reputable counterparts, however, the predatory journals exist primarily to pocket the publication fees, and there often is no real peer review. Authors publishing in predatory journals can benefit from having more publications to list on their résumés, but only if nobody looks closely at the publications themselves! The quality of the papers in predatory journals can be truly atrocious, and the papers are often blatantly plagiarized from other published research. These journals often do have very general titles, but the research published in them can be bad or even fictitious. It's a good idea to investigate who publishes a new journal that you haven't heard of before—if it's not connected with a major publisher or scientific society, it may not be trustworthy. Searching the World Wide Web for "predatory journals" will enable you to find published lists of known predatory journals.

FOR FURTHER STUDY AND DISCUSSION

1. Choose a volume of the *Philosophical Transactions* online, from the first two centuries of publication, and look at five different articles or letters. For each, answer the following questions:
   a. Is it a "letter" or an "article?" How did you decide?
   b. How many references does it make to previous work?
   c. How much detail is there on the experimental procedures or observations? Try to summarize, in a paragraph, the ways in which this older scientific literature differs from the scientific literature of today.

   Your instructor may assign you a particular year; in class, you can compare notes with other students who have been assigned other years to get an idea of how the scientific literature has evolved over the first 200 years of scientific journals.

2. Find a journal in your discipline that has been around for at least eighty years, and for which your university or college has online access for all of those years. Pick issues from the archive at ten-year intervals from 1930 through 1960. Just page through the articles and classify them as "not IMRAD," "sort-of IMRAD, but not strictly," and "strictly IMRAD." Graph the trends in the percentage of articles that fall into each of these categories as a function of time. This is best done as a group or class project, sharing the work of looking through the different issues of the journal.

3. Find a paper from a nineteenth-century scientific journal—well before the IMRAD format became popular. Read through the article and try to figure out what material would now go into an *Introduction, Methods, Results,* or *Discussion* section. You can try to do this at the paragraph level, but you may find that you need to go to the sentence or even phrase level to sort out the different components.

ADDITIONAL READING

Gross, A. G., Harmon, J. E. & Reidy, M. S. *Communicating Science. The Scientific Article from the 17th Century to the Present* (Parlor Press, Inc., 2001).

    Study of the form and rhetoric of the scientific literature is still in its early stages. Alan Gross has been a pioneer in this area, and this book is a good introduction.

Standage, T. *Writing on the Wall. Social Media, the First 2,000 Years* (Bloomsbury USA, 2014).

    This book is about a whole lot more than scientific journals but provides some interesting context for the founding of the *Philosophical Transactions*.

CHAPTER FIVE

ABSTRACTS COLLECTIONS AND DATABASES

LEARNING OBJECTIVES

After reading this chapter, you will be able to:

- explain the history of abstracting publications and how they led eventually to the databases that scientists depend on for finding literature
- identify the kinds of advantages and disadvantages that different databases offer, and choose the databases most relevant to your research needs
- implement search strategies that take advantage of Boolean operators and truncation operators ("wild cards") available in sophisticated databases

5.1 A BRIEF HISTORY OF ABSTRACTING AND INDEXING

In chapter 3, we made a brief hike into the landscape of the scientific literature. We focused there on the resources that were most helpful in getting started—general reference works and books—and how to find such things in a library using a catalog based on the

Dewey Decimal or Library of Congress system. But such a catalog isn't much use in finding individual articles in journals. For that, we need to use abstract collections and journal databases.

Abstracting was used to make articles more accessible even before the modern library cataloging systems were in place. Early in the nineteenth century, journals began to be published that consisted mostly of abstracts of articles published elsewhere. For example, in 1830, the *Pharmaceutisches Central-Blatt* was founded in Leipzig, Germany; it published abstracts of chemistry research published in other journals. Later renamed the *Chemisches Zentralblatt*, it continued publication until 1969. American chemist Arthur Noyes founded the *Review of American Chemical Research*, which also published abstracts. This journal became *Chemical Abstracts* in 1907, and the Chemical Abstracts Service continues to provide the most important worldwide search tools for the chemical literature today. *Biological Abstracts* was created out of the merger of *Abstracts of Bacteriology*, started in 1917, and *Botanical Abstracts*, started in 1919, both American publications. *Physics Abstracts* arose out of *Science Abstracts*, founded in 1898 in London. In these original abstracts journals, abstracts were typically written by *abstractors* who read the original publications, not by the authors of the papers. These abstractors were usually volunteers.

Abstracts that were written by the authors of the scientific papers themselves didn't start showing up in the most prominent science journals until the mid-twentieth century. The most prominent American chemistry journal, the *Journal of the American Chemical Society*, didn't regularly publish abstracts alongside articles until 1951; *Science* began regularly publishing abstracts with reports in 1959; and *Nature* didn't regularly include abstracts until 1967. The last volunteer abstractors for the Chemical Abstracts Service wrote in the 1960s. The shift to having authors write their own abstracts has obvious advantages for both the author and the abstracts journals; authors get more control over how their work is represented in an abstract, and the abstracts journals don't have to rely on volunteers or paid staff to produce abstracts. Some people, however, think professional or trained volunteer abstractors might do a better job than the authors in some cases. Condensing the essence of an article down to a paragraph or two is a difficult task in writing, and someone who is good at doing the science may not be the best at writing a summary of it.

One of the original reasons for the earliest abstracts journals was to give scientists access to the science published in journals in other nations and in other languages. In the nineteenth century, *Chemisches Zentralblatt* allowed German chemists to read, in German, about chemistry published in the *Quarterly Journal of the Chemical Society of London*, the *Comptes rendus de l'Académie des Sciences* of Paris, and the *Gazzetta Chimica Italiana*. Arthur Noyes's intention in founding the *Review of American Chemical Research* was to publicize the work of American chemists in Europe. In the days before airmail, photocopying, and the Internet, getting copies of articles that were published in foreign journals was quite a project. An abstract published in an abstracts journal allowed a scientist to get the gist of what a foreign article was about, and to decide whether to go through the work of getting the actual article.

Even when I was in graduate school, abstracts still served this purpose to some extent; even at a well-stocked academic library in the United States, there were likely to be journals that weren't available, and requesting these articles through Interlibrary Loan (ILL) would mean a wait of a week or two, as the article had to be photocopied and sent through the mail. The abstract was therefore important in helping scientists decide which articles to request. For scientists in poorer nations, who were likely to have access to many fewer journals and no Interlibrary Loan, it was all the more important to get an idea of what an article was about before trying to get a copy of it. Scientists in such countries had to request reprints of articles directly from the author.

Now, with global Internet access, getting papers from all over the globe is not as much work, and doesn't take nearly as long. Even though it may still be necessary to use ILL, electronic transmission of the articles from other libraries means that it only takes a day or two to get an article from a distant library. So, while abstracts still help scientists figure out whether to get articles, that's no longer their primary purpose. Now they are more likely to be informing people whether it's worth *reading* a paper, rather than whether it's worth sending away for it.

Like the abstracts themselves, the abstracts journals also have evolved to serve other purposes. The large number of abstracts published meant that, even from the beginning of the twentieth century and earlier, the abstracts had to be classified and indexed. Gradually, the indexes became far more important than the abstracts collections themselves. With computerization, increasingly sophisticated means of searching the indexes based on keywords, authors, chemical substances, and even related chemical structures and reactions became possible. Now, scientists think of the *Chemical Abstracts Service* far more as a source of the online search tools in *SciFinder* than as a source of abstracts; the abstracts themselves can always be found at an individual journal's website.

Not all guides to finding scientific articles started with abstracts; *Index Medicus*, the printed precursor to the online medical literature services *MEDLINE* and *PubMed*, was founded in 1879 as a bibliographic index. Abstracts weren't included; keywords or words in titles could be used to find out in what journals relevant articles had been published and when, and who the author was, but there wasn't any information beyond the title to indicate what was in the articles. More recent indexes to the scientific journal literature also avoided including abstracts before the age of cheap computer storage; *Science Citation Index (SCI)*, which we will discuss more in the next chapter, was focused from the beginning on the citations or references in journal articles rather than the abstracts. Now both *PubMed* and *Web of Science*, the modern Web version of *SCI*, display abstracts, as these are easily retrieved from publishers electronically and added to the databases.

As journals moved more and more to online access, journal publishers also did their own sort of indexing. The websites of many major publishers have a search engine that will search the journals of that publisher for keywords, authors, and other search terms. These are sometimes the most accessible search tools available for journal articles, but you need to be aware that their coverage is limited to a particular publisher's journals, sometimes called a *journal collection*.

The innovations in technology that have occurred since the first abstracting journals were published have changed indexing and searching. When indexes were all in print, the process of indexing tended to rely on *controlled vocabulary*, like that used by typical library cataloging systems. Indexers would examine an item in a journal and assign it index terms that were part of the controlled vocabulary that the indexing service had established. Searching for articles meant finding the controlled vocabulary terms that were most relevant to your information need and looking those terms up in a subject index. Sometimes this required a little trial and error, if the terms you thought were the most descriptive turned out not to be in the controlled vocabulary. Print indexes were not just limited to subject terms, however. The old printed form of *Chemical Abstracts*, for example, included a General Subject Index; a Chemical Substances Index, where chemicals were referred to by name; a Formula Index, where chemicals were referred to by formula; and an Author Index. Each of these indexes had its own controlled vocabulary or rules for entries.

As computer searching became possible, controlled vocabularies and separate indexes became less important. It became possible to have a *database* with different *fields* for the full text of the title, the abstract, the author, and controlled vocabulary subject terms. Users could then search in some or all of these fields in the database for *keywords* that might appear in the title or abstract for articles they might be interested in. Some databases, such as *Chemical Abstracts Online (CAOnline)*, gave the user a lot of control over just how searching was done. Users could specify that the search be confined to particular fields, or cover all fields. Certain terms might only be searched in one field, such as the registry number of chemicals mentioned in an article, and the results of this search then combined with a general search of the title, abstract, and subject headings for particular keywords. This kind of searching could be very powerful—if you were properly trained in how to use all the tools and had plenty of practice.

With the coming of the World Wide Web and greater computing power, search engines such as Google changed the way people thought about searching for information. *Full text searching* became more popular. A search engine for the entire Web could not rely on indexers to assign controlled vocabulary subject terms to billions of

Web pages. Instead, automated *Web crawlers* retrieved the full text and all links from Web pages; algorithms then analyzed the text, the links between pages, and other factors to make a best guess of which pages would be the most relevant when a user entered particular keywords or a question. The algorithms used in such search engines are proprietary, so you don't know exactly how your search is being processed, but as long as users are happy with the results, people continue using these search engines.

The ease of searching the Web using Google and similar search engines has changed how popular science databases now function. In *SciFinder*, for example, the Web-based successor to *CAOnline*, the search entry page now encourages natural language queries such as, "The effect of antibiotic residues on dairy products," and then decides for you what search terms to use—in this case, "antibiotic residues" and "dairy products." No longer do you have to specify particular Boolean operators (see section 5.3); the search engine will assume a number of possibilities and give you the result for each. For someone used to the precision and control of the old way of searching, it can be a bit frustrating, as it's not always clear exactly how the search engine is interpreting your query or where it's looking for matches in the database. But the ease of use is probably welcome to those who have grown up searching on Google.

5.2 INVESTIGATING DATABASES

There are many different databases that allow you to search the periodical literature—journals and magazines—in specific fields. I've already mentioned some of the most important ones in the history section above: *Chemical Abstracts*, *Biological Abstracts*, *Physics Abstracts*, *PubMed*, and *Web of Science*. And I also mentioned that many journal publishers have their own databases that search the literature that occurs within their own journals. Which ones you use will depend on what field you're trying to find articles in, what time period you want to search, and what is available to you at your library. Some of the computerized databases can be quite expensive for libraries to get subscriptions to, and the print versions of these can take up a lot of space—though most aren't even printed anymore.

It doesn't make sense for me to give you specific instructions for searching particular databases, because not everybody is interested in, or has access to, the same ones. Also, they tend to change their user interfaces frequently as they adapt to changes in technology. It's better to talk about general strategies for using such databases; your professor or librarian can give you the details on what is available to you now and how to use these resources.

One of the first things you ought to do is figure out the basics of the database. Is it a very general science database, or does it focus on a particular field or closely grouped fields? If you really want a definitive answer to this, you should find out just which journals the database covers—although in some cases, this may take more time than it's worth. It may be better just to imagine what the limitations might be. If you are looking for a topic in chemistry, *Chemical Abstracts* is the best place to go; if you want to find more biological information, try *Biological Abstracts*. But don't assume that you can *only* get information from what would seem to be the *best* database. You can find a lot of chemistry in *Biological Abstracts*, and a lot of biology in *Chemical Abstracts*. A lot of both chemistry and biology is covered by *PubMed*. If your access to databases is limited, use what you can, find what you can, and use it as best you can, making good use of the technique of tracing references, which we describe in the next chapter.

Although there is a lot of overlap between databases, keep the limitations in mind—you may be missing something if you aren't using the most appropriate database. This is especially important when you want to do an *exhaustive search*. An exhaustive search is what you want to do if you want to claim that you're the first researcher ever to discover something or invent a new method of doing something. If you're a chemist who wants to claim that you're the first person to synthesize a given molecule, it doesn't make sense to make that claim after just searching *PubMed* and *Biological Abstracts*. You should search *Chemical Abstracts*, and it would be good to do the search with someone who really knows how to use the resource.

Some databases contain more than journal articles; *Chemical Abstracts*, for example, also includes patents, conference proceedings, and more. Individual publisher databases may also include books issued by that publisher. This is also something that you will want to be aware of when using a database.

Another thing you want to determine is the time coverage of the database. *Chemical Abstracts* goes back to 1879; the database of *Web of Science*, which we discuss more in the next chapter, covers the literature for more than 100 years, but not all libraries have access to that entire time period. This is because the publishers of *Web of Science* charge libraries more for access to the full database, and *Web of Science* is already a rather expensive database. Publisher journal collections usually focus on the last few decades, as they only have electronic versions of their journals for this amount of time, and putting older issues online requires scanning the documents. Scientific societies, being less profit-driven, have more of an interest in doing this; the American Chemical Society and the Royal Society both have scanned copies of their oldest journals available online.

5.3 IMPLEMENTING A SEARCH

In this age, most searches will start with keyword searching rather than controlled vocabulary search terms. Choosing keywords or search terms is something of an art, and you want to think about where words would occur in an article. Different parts of an article may be searched by a database; words in a title will almost always be searched, but some will also search the abstract, or even the whole text of the article. Author-generated keywords published alongside the articles will be included in some cases. If you do a little research, you can find out what is searched by a particular database. Knowing whether database searches look at just the title, or the title and the abstract, however, doesn't necessarily help you come up with better search terms, a task that is often the hardest part of doing an effective search.

In choosing search terms, experience and imagination are once again very useful. For example, if we were researching laundry detergents in *Chemical Abstracts*, "laundry" and "detergents" may not be the most useful search terms. As you might imagine, PhD-level chemists publishing in the scholarly chemical literature are more likely to focus on the components of detergents and their interactions with particular materials, so "surfactants" and "textiles" may be more appropriate terms to use. Remember that getting this more advanced vocabulary was one of the things we accomplished in chapter 3, when we were taking a look at the general literature landscape.

It is also important to know how, and when, to combine search terms. Many databases accept *operators* from Boolean logic to help you narrow, or broaden, your search. If you use the Boolean operator "and," you will make your search more restrictive. "Laundry and detergent" will only return results that contain both. "Or," on the other hand, broadens a search; "detergent or surfactant" would find both articles about surfactants that never mention their use in detergents and articles about detergents that don't ever mention surfactants. Combining both strategies is sometimes necessary; if you were getting too many articles about dishwashing detergents with that last search, you might want to do something more complex like "(laundry or textiles) and (detergent or surfactant)." This is not possible in all search engines, or it may not be possible to do in one step, but it's worth thinking about how to do the equivalent of this kind of search in whatever database you have handy.

Most search engines have the "and" and "or" Boolean operators; some also have the "not" operator, which can be especially useful in some bigger databases. In *CAOnline*, the older online tool for searching *Chemical Abstracts*, the last search formulation above would find a lot of patents issued for detergent formulations or surfactants useful in the cleaning of laundry and textiles, because the database includes patents. If we were interested in eliminating all of the patents, we could add a "not patent/dt" to our search above to get rid of all the items listed as patents in the document type (dt) field. *SciFinder*, the successor to *CAOnline*, enables you to do something similar with a "Refine" option.

Two words written together without an operator between them are sometimes interpreted as a single search term; in some databases, however, quotation marks around the two words are necessary in order for the search engine to retrieve only articles where both of those words are adjacent in the entered order.

Truncation operators are important to know about, too. These are often called "wild cards." These enable you to catch all variants of a root word; in many databases, "symmetr*" would find articles containing the words symmetry, symmetries, and symmetrical. Not all databases use the same symbols for truncation operators, and some allow several different kinds of truncation operators, so it's a good idea to investigate how your database uses them.

Just because keyword searching has become more and more powerful, you shouldn't ignore the use of controlled vocabulary subject headings. These subject headings are assigned thoughtfully by indexers who know a field well, and you can make use of their expertise in finding the articles that are the most useful for your purposes. In *PubMed*, for example, you can find the controlled vocabulary subject headings for any given article. If you go to the detailed record for a suitable article you can find a link to *Medical Subject Heading*, or *MeSH*, terms used to classify that article. Clicking on the *MeSH* term that you think will be most useful to you can lead to lots of other articles about the same topic. There are also ways of going directly to *MeSH* terms from a keyword search; the search engine uses a thesaurus to match up your keyword with the most appropriate controlled-vocabulary subject heading.

An important part of doing a search through the periodical literature is coming up with a manageable list of references. You can't read through 240 scientific articles in the course of a semester. I typically aim for 10 to 20 articles in a final search result. Some of these may be in obscure foreign journals that are really hard to access. Others will turn out to be about something rather different from what we're interested in. That should leave at least 3 to 5 articles that are worth looking at seriously, if not reading from beginning to end. One reason I don't aim for more results in the database search is that the database is not the final step in the literature search. Each of the 3 to 5 good articles that result will have cited references, and they will be cited by other articles. Using the cited references in an article, and using other search tools that allow us to find articles that have cited that article, can lead us to many, many more papers that may be relevant to our research. The next chapter describes how to do that.

FOR FURTHER STUDY AND DISCUSSION

1. Explore several of the databases that your library has available for searching the scientific literature. Answer the following questions: Who owns and maintains the database? How many journals are covered by the database, and what's the best way of describing this group of journals? Are the journals all from a single publisher? How far back does the database go? What sort of Boolean operators or truncation operators can you use in searching?

2. Pick some words that would be appropriate for searching databases for information relevant to a class project. Figure out how to best use Boolean operators to capture just the articles you want, without getting too many that are irrelevant. Try this search in one of the databases available to you. Vary the search terms and operators and see how these changes affect your results. It is possible to make a search too restrictive, so you should also check to see whether your restrictive search is leaving out good articles. You can do this by making the search broader and seeing whether there are relevant results that were excluded by your more restrictive search.

3. Compare your results using the same search terms in several different databases and *Google Scholar*. Which databases give you the most relevant results? Are there ways to fine-tune the search strategy in different databases so that the results from the databases are more similar? If not, why not?

ADDITIONAL READING

Lancaster, F. *Indexing and Abstracting in Theory and Practice* (University of Illinois Graduate School of Library and Information Science, 2003).

Although this book is written more for library professionals than for the casual reader, it gives wonderful insights into just how hard it can be to connect information users with the information they need. Indexing and abstracting are indeed fascinating arts!

Subramanyam, K. *Scientific and Technical Information Resources* (Marcel Dekker, 1981).

Subramanyam covers indexing and abstracting services near the end of his book.

CHAPTER SIX

USING CITED REFERENCES—BACKWARD AND FORWARD

LEARNING OBJECTIVES

After reading this chapter, you will be able to:

- identify the different roles cited references can play: Background, Methods, and Other Research in the Field
- classify references as falling into one of these categories
- predict what you will find in cited references based on the context in which they were cited
- explain the value of following references back several generations to get a sense of the history of a field of investigation
- identify databases that allow you to look forward to other articles that have cited an older article

6.1 THE IMPORTANCE OF CITED REFERENCES

You should be familiar with the need to cite your sources when writing a paper for school. The reasons you have heard for this are probably along the lines of "You need to show me that you read at least three sources before writing this paper"; "You need to give credit to the sources where you got your information"; or "You need to show that the sources of your information are reliable." Scientific papers cite previously published journal articles and books; these cited references share some of the same goals as the citations you make in schoolwork, but they can also be used to expand your knowledge of the literature landscape.

Another thing that may have been emphasized in your previous schooling is that your references or citations need to be in a specific format. I'm *not* going to emphasize that in this chapter, because disciplines, sub-disciplines, and publishers all have their own styles. For example, the *Journal of the American Chemical Society* has a different style than *Biochemistry*, even though they are both published by the American Chemical Society; other biochemistry journals, such as *The Journal of Biological Chemistry*, have a slightly different style than *Biochemistry*. In this book, we've used a format based on that used by *Nature*, because it's an international journal that publishes work in all sorts of different sciences. The important thing is to be flexible and use whatever format your instructor, or the journal you want to publish in, wants. This is made easier now by the many brands of *reference management software* that can store all the data in a standardized format and then just put it into whatever format you want or need for a given paper or assignment.

References in scientific articles do all the things I listed in the first paragraph, but they play a few other roles. One of these roles is to shorten scientific papers. As I discussed in chapter 4, there has been a huge increase in the amount of science being published since World War II. Journals have a hard time publishing the flood of papers, and extra pages take up space and cost money. Although the World Wide Web may help alleviate some of the cost of longer articles, the current situation is that journals still require papers to be as short as possible. The editors of journals don't want every paper in a given field to cover all the background to a particular problem, or all the methods that everybody who investigates that problem uses. Therefore, papers will typically have references in the *Introduction* directing readers to articles that have more background on a research field, as well as references in the *Methods* or *Experimental Section* to articles that give the details of particular methods used in the research. This is one reason papers can be hard to read; if you're not already familiar with the background and standard methods, you'll have to look up the referenced papers and educate yourself about these things.

Another thing that cited references do is to help connect an individual research project in the context of what other people are doing in a particular research field. As we will see in chapter 7, where we look at how to identify arguments in scientific papers, individual papers rarely encompass complete answers to important scientific questions. Many lines of experimental, observational, and computational evidence must be assembled before scientists feel that they really know what is going on in a particular system or phenomenon, and the whole picture can only be seen by looking at many papers from many different authors. We will see that *Introductions* of papers often refer not only to papers that give more general background but also to individual studies that have made incremental contributions to the overall field.

6.2 LOOKING BACKWARD

Given the importance of both of these purposes—shortening the paper and giving the research context—it is obvious that one of the best ways to get a feel for the literature landscape in a given field is to start looking up papers and books that have been cited by a paper. Of course, we have to keep in mind that a paper can only have cited work that has already been published, which is why we call this section "looking backward." We are looking at what came before the paper we are reading.

Because many journal articles have dozens of references, you need to be smart about which cited references you look up. People doing a PhD thesis in a field may end up reading *many* of the papers cited by the central articles in that field. But if you're *not* doing a PhD and don't have lots of time, you're not going to be able to read all the papers cited by all the papers that are cited by all the papers cited in that first paper. Given that the average experimental paper in the twentieth century had twenty-six references,[1] reading all the papers referenced in a paper, all the papers referenced in those papers, and all the papers referenced in *those* papers could get you as many as 26^3 papers, or 17,576 papers! That's probably a substantial overestimate of the actual number of papers, since many of the papers in a field will be citing some papers in common. But even if 90 percent of the references are repeats, you'll still have 1,758 papers to read, and you aren't going to have time to read all of those, in most cases.

The key to using cited references is to get good at guessing the contents of a reference. You can do this by looking at the context in which it is cited, the title (if given in the journal's standard reference format), and the journal in which it occurs. I'm going to show you a number of examples of how you can do this, focusing on getting clues from the context in which it is cited. These examples are taken from a wide variety of journals, and the reference format is not the same for all the journals. Also, you may find the science in the following examples unfamiliar. Don't worry about that; some of it is unfamiliar to me as well! I used these examples on purpose to illustrate how you can get a sense of what you will find in a reference just from the way it is cited. I usually categorize references into one of three types: Background References, Method References, and Other Research in the Field:

1. **Background References:** These are often references to a review paper, a theoretical paper, or even a book. These references often are placed after a very general statement in the *Introduction*. Here are some examples:
   a. "Secondary structure formation has been proposed as one of the early steps in the folding pathway by which a polypeptide chain folds into its native three-dimensional structure.[1-3]" A look at the references 1 to 3 shows that they are found in *Annu. Rev. Biochem.*, and *Curr. Opin. Struct. Biol.* Both of these sound like review journals; *Annu. Rev. Biochem.* is *Annual Review of Biochemistry*, and *Curr. Opin. Struct. Biol.* is *Current Opinion in Structural Biology*. Because these are all review articles, any one of them—or all of them—would be good places to go to find background on the problem examined in the paper. (Finding titles from journal abbreviations isn't magic; most Web search engines will tell you right away what the abbreviation stands for.)[2]

b. "Over a century of research into the origin of turbulence in wall-bounded shear flows has resulted in a puzzling picture in which the turbulence appears in a variety of different states competing with laminar background flow.[1-6]" All six of these might potentially be Background References, but a look at the titles of the articles referenced gives you some clues as to which would be most useful if you're looking for background. The title of one of them begins "An experimental investigation of the circumstances...." You don't have to read any farther to see that this is not a review paper or a theoretical paper—the "experimental investigation" tells you that. Most likely it's an *example* of the sort of system that shows the origin of turbulence in wall-bounded shear flows, so it probably doesn't qualify as a General Background reference.

There are three more references that look like experimental papers that just give examples, judging from their titles. But one of the first six references has a more general title, and it looks as though it comes from a book that may be an edited volume or conference proceedings: "Coles, D. Interfaces and intermittency in turbulent shear flow. In *Mécanique de la Turbulence* 229 (Editions du Centre National de la Recherche Scientifique, 1962.)" One clue that it is a book is the word "In," which tells us that this is just a chapter of a book or a paper in a conference proceedings or review collection. This is probably a better background reference than the first, but the chapter may be in French. Then there is another reference from *Annu. Rev. Fluid Mech.* in 2011. This is a good find, because it's more recent than 1962, it's not in French, and it's from a review journal (*Annual Reviews in Fluid Mechanics*). In this case this is probably the one you want to go looking for, unless you are good at reading scientific French. Sometimes you may not want the most recent general background paper; an older paper will help us get a wider variety of papers when we "look forward," as explained in the next section.[3]

2. **Method References:** These are most often found in the *Methods* or *Experimental Section* and will be found exactly where you'd expect more details on a given method. Here are some examples:

   a. "Unlabeled m^6dGTP (Foote *et al.*, 1980) and [8-^{3}H]m^6dGTP (5.8 Ci/mmol) (Foote *et al.*, 1983) were prepared as previously described." That's a pretty straightforward indication that the articles by Foote *et al.* are where the methods were previously described. However, we can glean more from these citations. The authors of the article with the above quotation are Snow, E. T., Foote, R. S., & Mitra, S. The two "Foote *et al.*" articles have authors "Foote, R. S., Mitra, S., & Pal, B. C." and "Foote, R. S., Pal, B. C., & Mitra, S." All three of these articles—the two that are being referenced and the one in which the references are found—have authors in common. This one laboratory most likely has developed the techniques and used them in a number of studies.

   b. "Poly(dA-dT) was synthesized in a primed reaction with *Escherichia coli* DNA polymerase I (Radding *et al.*, 1962)." This is from the same paragraph as example *a* above, and is also pretty straightforward. But here, the authors are using a much older, probably well-established method; there are no authors in common between the citing paper and the paper cited, so the method was probably developed in a different laboratory.[4]

3. **Other Research in the Field:** These can show up almost anywhere in a paper. They can be in the *Introduction*, where the author uses them to put the present research in context of what has been done. They can be in the *Results*, where an author might compare his or her results with other results that have been reported in the literature. They can show up in the *Discussion*, where the author points out agreement or disagreement with other authors' discussions (interpretations) of their results. Here's an example from a results section:

   a. "Although examples of Suzuki coupling to form very hindered biphenyls from aryl iodides or bromides have been reported,[6b,15] reactions of this type are often problematic.[1,6b,15] In some cases the use of certain bases (TlOH,[6b] Ba(OH)$_2$,[15a] or K_3PO_4[15a]), or solvent combinations (e.g., toluene/water/ethanol, 3/3/1)[15e] have been reported to give improved results, although the generality of these protocols is not clear." There's a lot going on in this passage, but we can make some pretty good guesses as to what we are going to find in the different references. References 6b and 15 should have examples of the formation of hindered biphenyls from aryl iodides or bromides. The appearance of reference 1 at the end of the first sentence suggests that the problems of this kind of reaction may have been discussed in a Background Reference, which are often the first references cited. Sure enough, reference 1 turns out to be by Suzuki himself (after whom the Suzuki coupling is named) and appears to be a chapter in a book on a variety of cross-coupling reactions.

In the second sentence quoted, you can see that reference 15 must actually consist of a variety of references stretching at least from 15a to 15e. Not all journals allow or encourage this, but it's rather common in the *Journal of the American Chemical Society*. Without even looking at the references, you can tell that reference 15a did experiments with $Ba(OH)_2$ and K_3PO_4, and that reference 15e explored reactions in solvent mixtures.[5]

Deciding on which references to pursue requires more than just classifying the references, though. In fact, you shouldn't go through *all* the references in a paper and classify them according to the scheme above. That's a waste of time. Instead, decide which category of references you're interested in first. If you're fairly new to an area of research, you probably want to look for background references. If you're familiar with the field but want to see what's been done lately, look at recent examples of other research in the field. If you want to understand what a particular laboratory or researcher has been up to, find examples of other research in the field that shares authors with the paper you're looking at. If you just want to find out how to do experiments in the area of research, look at the *Methods* or *Experimental Section* and find method references.

It is a bit unrealistic to focus here only on the references, without thinking about the actual task of reading the paper itself. We cover that in a later chapter, and there we mention that being willing to look up references is part of reading a paper successfully. Often, the things that are most confusing about a paper may be the things you need to look up the references for, just to understand the paper. One point of this chapter is that, even without reading a paper much, you can use cited references in a paper as a source of information of what had been previously published in the field.

There is one story that I always tell my students about using references from a paper. When I was an undergraduate, I had to do three "junior papers" with three different chemistry professors. One of them handed me a recent paper about organic stereochemistry. In that paper were the results of theoretical calculations of energies for some sterically hindered molecules—molecules where various parts of the molecule got in the way of other parts. I didn't understand how they did these calculations. But there was a method reference, and I looked it up. This second paper I looked up didn't help me much, but it gave me a little bit of an idea of how these computer calculations worked. So, I looked up a reference in that second paper, which was, of course, even older, and I learned a bit more. I don't remember how many times I repeated this, but eventually I got back to rather old references where the ideas of computing energies of different molecules' shapes by *molecular mechanics* were first introduced—back before computers were readily available to do the calculations. There, the basic ideas were very clearly explained, because they were new ideas! I could then follow how, in later papers, these ideas were increasingly implemented on computers and how more advanced techniques of computer optimization were tried. After having gone so far back, I felt that I had some perspective on those calculations in the original paper that the professor had given me, and I could use these to critically examine some of the interpretations of the calculations in the paper. The point of this story is that you can learn a lot from references, but you have to be willing to keeping going back to older and older references until you find the roots of an idea or technique.

6.3 THE LIMITATIONS OF LOOKING BACKWARD, AND THE NEED TO LOOK FORWARD

So far, we have only been *looking backward*, finding articles that were published *before* a particular article. If you have a fairly recent article, this may not seem to be that much of a problem, as there have probably been few articles on the topic published since that article. But often you will want to *look forward* as well. Sometimes, the first paper you start with will not be very recent, and you'll want to catch up on what's been done in a field since that paper was published. Also, older articles that you found in the process of looking backward may have been cited by other more recent articles that you don't yet know about. This is illustrated in figure 6.1; the articles to the right of the dashed line are not cited by the original paper or other papers you found by looking backward. They do, however, refer back to an older background reference and other papers that cited that reference. These may be papers that take a very different approach to the same general problem originally outlined or reviewed in the oldest general background paper, and therefore may be very valuable in broadening your view of how other scientists have approached a problem.

Science Citation Index/Web of Science. The problem of looking forward in the scientific literature was first tackled by Eugene Garfield in 1955, who wrote a paper in *Science* outlining the many advantages of a *citation index* that keeps track of, and regularly publishes, a list of the articles that cite a previously published article.[6] (It is interesting to note that the legal profession had such a citation index long before science did; Garfield in many ways modeled his system after *Shepard's Citations*, the legal research tool established in 1873.) Among the advantages that Garfield outlined was the possibility of allowing a reader to find out how an older paper had been criticized or even refuted by subsequent researchers. Such an index would also be a great help to historians of science who wished to trace how an idea had shaped the thinking and research of subsequent scientists. One use of such an index that has become somewhat controversial is its use in evaluating the quality of scientific papers; the idea is that papers that are cited many times are more

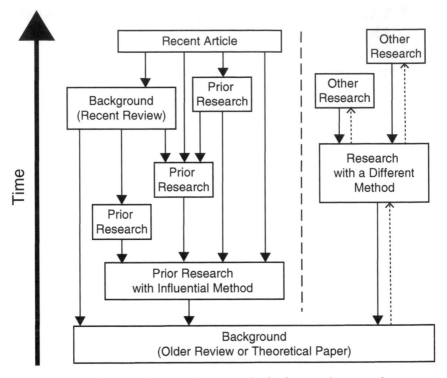

Figure 6.1. A schematic showing the network of references between the "Recent Article" you might start with and other articles. The articles to the right of the dashed line cannot be accessed by looking backward through the citations in the original article or the articles it references; they can only be accessed by *looking forward* using one of the tools described in this section (as represented by the dotted arrows).

likely to be important, high-quality science. This is controversial because people worry that good work may not be cited for reasons that have less to do with its quality than with politics and publicity, and the rankings of how often a paper is cited can be used for deciding who gets tenure at prestigious universities or influencing who gets research grants.

In 1960, Eugene Garfield founded the Institute for Scientific Information (ISI), which, after a number of trial projects of more limited scope, eventually ended up publishing the *Science Citation Index (SCI)* in 1963. Key to the success of the ISI's endeavors was the increasing accessibility of computers. Although the computers were helpful in assembling the *Index*, the output was in printed volumes, and continued to be accessed mostly through printed volumes into the early 1990s. The twenty-first century brought a transition to a World Wide Web product that is much easier to use, called *Web of Science*. Its name is a bit of a play on words; it reflects not only the fact that it is accessible via the World Wide Web but also that it is good at tracing the web of connections in the scientific literature shown in figure 6.1. Although the original idea for the *SCI* was based on citation searching, visitors to *Web of Science* are now typically greeted with a *Basic Search* option. To find articles that reference an older article, the sort of search that *Web of Science* does best, it is best to select a *Cited Reference Search*.

I won't go into the details of how you do such a search, because there is no guarantee that the Web interface will work the same way through time—there have already been a lot of improvements in the years I have been using it. Your librarian or instructor should give you a more relevant tutorial on how to use this resource if your campus has it available. Whether or not they have it available depends a lot on how much money your university or college has, as a subscription to this resource is quite expensive.

Alternatives to Web of Science. When Eugene Garfield had his idea for the *Science Citation Index*, one of the big issues was how to get the information that was in *print* to some form where it could be manipulated by *computers*. By the 1990s, that wasn't really the issue anymore. With the coming of the World Wide Web, the papers that previously were only public in print started showing up as digital files on the Web. This raised the possibility that a Web crawler—an automated program that systematically browses the entire Web—could find all the scholarly or scientific papers or books on the Web, harvest the citations, and create its own citation index.

Google Scholar is the most popular of the automatically generated citation indexes; it was first rolled out in 2004. Unlike *Web of Science, Google Scholar* doesn't have a fixed list of journals that it indexes. This potentially allows it to find citations that *Web of Science* might miss, from books, theses, and other sources not indexed by *Web of Science*. But the materials available to the Web crawler will depend on whether journal publishers allow the Web crawler to access their journal, so there's no guarantee that all citations will be found. The nice thing about *Google Scholar*, however, is that it is free to anyone who has Web access. One thing you should be careful about, however: the default ordering for results in *Google Scholar* is based on some algorithm that tries to pick out the "most important" results. This algorithm is proprietary (we don't really know how it works), and people have shown that you can fool the algorithm into putting your paper at the top of the list!

Scopus is another competitor to *Web of Science* that emerged in 2004, from the journal publisher Elsevier. Like *Web of Science*, it has a fixed list of journals that it indexes, but this list is larger than that of other databases, and the literature surveyed includes books, conference proceedings, and patent databases. Also, like *Web of Science*, it is a commercial product that costs money to get a subscription to. While the *Web of Science* database reached further into the past than the *Scopus* database when the two were compared in 2006,[7] it appears that *Scopus* is continuing to expand its coverage to older literature.

Disciplinary search engines such as *SciFinder*, the Web-based successor to *Chemical Abstracts*, have also incorporated cited reference searching tools into their product.

There have been a lot of people who have tried to determine which of these various tools gives the most *exhaustive* coverage of the literature—in other words, which among them finds *every* citation of a given work. The results one gets depend a lot on which discipline, and even which sub-discipline, of science is the topic of the search. Sometimes, doing an exhaustive search is necessary, especially if you want to claim that "nobody has discovered this before." Using a combination of tools is ultimately the best way to do such a search.

For most undergraduate students doing research, however, an exhaustive search is usually not your end goal. If you find a thread that seems important, going forward *or* backward, you should pursue it. If you see a lot of citations or articles that seem peripheral to what you're really interested in, ignore them. It is important to have an idea in your head of just what sort of information you are most interested in.

FOR FURTHER STUDY AND DISCUSSION

1. Pick out an article that is relevant to your at-home research project or a library research project in a different class. Write down the bibliographic information for this reference (author, journal, publication date, volume, page) in the format that your instructor specifies. This article should be a primary source, with original research being presented.

   a. From this article, find three references that illustrate the three different types of references: *Background*, *Method*, and *Other Research in the Field*. Choose these references based on what you think will be most helpful to you in your research, *and* which occur in journals that you have access to through your library. If you have to get an article through Interlibrary Loan, that's okay, but just keep in mind that this may take some extra time. Write down the bibliographic information for these references in whatever format your instructor specifies. Based on the context in which the reference was made, make a prediction about what you will find for each reference.

   b. Retrieve a copy of these three articles electronically or in print. Report whether your predictions about what was in the articles were correct. Also, describe some aspect of the research that you *didn't* expect to find in the article; part of the value of tracing down references is learning unexpected things! If the article is available at your school using the Web, then you don't have to include a copy of the article; however, if you got it through ILL, make sure to send a copy to your instructor.

   c. For *one* of these three articles—the one that leaves you the most puzzled or curious—try to find a reference in that article that you think will best help solve the puzzle or satisfy your curiosity. Classify this reference as *Background*, *Method*, or *Other Research*, predict what you will find in this article, and then go get the article. As before, report whether your predictions were correct, and describe some aspect of the research that you didn't expect to find.

2. Choose an article that is at least five years old and that you have some interest in. Older papers found in the previous exercise would be excellent candidates for this exercise.

   a. Using *Web of Science*, do a *Cited Reference Search* on the article to find more recent articles that have cited the article you chose. If your article does not yield at least three citations in the cited reference search, choose an earlier article from the references in the article you chose. Repeat until you successfully get at least three citations.

   b. Find the actual article for one of the citations. Describe why this article referenced the original article you chose in step *a*.

   c. Try putting the article you chose into *Google Scholar* and see how many "cited by" references you get. If the number of references is different for your *Google Scholar* and *Web of Science* searches, see if you can find possible reasons for the discrepancies between the two lists.

ADDITIONAL READING

Garfield, E. Citation indexes for science: a new dimension in documentation through association of ideas. *Science* **122**, 108–111 (1955).

   This is the article that started the idea of creating indexes that look forward.

CHAPTER SEVEN

READING A SCIENTIFIC PAPER

LEARNING OBJECTIVES

After reading this chapter, you will be able to:

- identify reasons why scientific papers are hard to read
- identify strategies to get past the difficulties of unfamiliar vocabulary, concepts, and methods
- classify figures and tables as being raw data, interpreted data, or explanatory figures
- extract information from figures and tables, and use the figures and tables to come up with questions about the research
- explain the difference between a local argument and a larger argument
- identify the parts of a paper most likely to contain the premises and conclusions for both the local and larger arguments
- give examples of critical questions that might be used to argue against a paper's conclusions

7.1 WHY IS IT SO HARD?

At some point in your career as a scientist—maybe last semester, next week, or a few years from now—a mentor may hand you a scientific paper (it may be an "Article," a "Report," a "Letter," or a "Communication") and say something to the effect of, "Take a look at this." That scientist would be thrilled and impressed if you came back the next day not only having *read* and *understood* the paper but having grasped the essential *argument* of the paper, formulated a *critique* of the paper and its argument, and come up with a *research plan* for expanding upon the research reported in the paper. That doesn't always happen, of course. It's actually a pretty difficult set of tasks, and lots of times younger scientists don't have the experience to pull it off, at least not on that timescale. But I hope that I can give you some pointers on how to make this ideal outcome a little more likely.

We're not going to cover the research plan part of the above scenario in this chapter, but we will be looking at how to read and understand a scientific paper and extract its main arguments. You may think you already know how to read, but scientific papers aren't things that just any literate person can read. Many undergraduate students, graduate students, and even professional scientists will admit, if you press them on it, that there have been more than a few scientific papers that they have struggled with.

And this is often the fault of neither the reader nor the writer—it is just a consequence of the nature of science writing. Hundreds of thousands of scientific papers get published every year. This flood of information has led most scientific journals to require or recommend a rather standardized organization and format for scientific papers. This standardized format means that one of the oldest and most readily understood forms of human communication—chronological storytelling—isn't used in a scientific paper.

Most research happens in a sort of looping, spiral fashion. You have an idea, you do an experiment, you get a result, which leads to a new idea, and a new experiment, and a new result; the cycle repeats many times. But this chronological story of how the research was actually done isn't communicated in a scientific paper. The standard

Introduction, Methods, Results, and *Discussion* format means that all the experiments go in one place and all the results in another, while the ideas may be split between the *Introduction* and the *Discussion.* The research might be easier to understand as a chronological story, but storytelling would result in articles that are too long (and maybe too boring) for a busy scientist to read. Scientists need to be able to find all the results or all the experimental techniques in one place, depending on what they are looking for.

The crushing wave of too many scientific articles has also led journals to insist on brevity even within the standard format. The author of an article isn't expected to define or explain terms that some people outside their specialty don't know—they use the specialized language of their discipline because it saves space. That's why even experienced scientists are sometimes stumped by articles that are outside their areas of expertise—they may not even understand the words in the title! The drive toward brevity also means that authors will typically not repeat descriptions of methods or theories that others have published previously, even if such descriptions would make it easier for readers to understand the article without a trip to the computer or library. You may be able to take a novel to your favorite comfy chair, read for a few hours, and understand most of what you read. That's rarely possible with a pile of photocopied scientific articles. You're always jumping out of the chair to look up a specialized term or a reference. I used to take articles on the train with me to read when commuting to and from work, but it was only a ten-minute trip. That's about as much time as I could profitably spend before I needed to go look up some background information in another paper or book.

If the scientific literature has become a highly structured and condensed code that efficiently communicates complicated ideas between highly trained specialists, is there any hope for an undergraduate student to get anything out of reading a typical paper in the literature? Yes, there is, but the undergraduate must first abandon the wish to read it from beginning to end while reclining in a comfy chair with a nice beverage. The undergraduate must also abandon the idea that they will understand every nuance of the reported research, which may have taken a bunch of scientists a few years—or at least months—to put together. The undergraduate has to approach the paper as a scientist would, with realistic expectations of what they will get out of the paper and not a lot of emotional baggage about what they *should* get out of the paper.

7.2 HINTS FOR TAKING A FIRST LOOK AT A SCIENTIFIC PAPER

I'm calling these "hints" because not everyone will agree that this is how students should be reading a scientific paper! Nevertheless, I think that these hints can be helpful when you don't have a lot of background in a research field and the authors are writing for people who already have a detailed knowledge of the field.

Hint Number One: Don't start at the beginning and read it straight through. Usually the first thing you see in a journal article is the abstract. As a result, people think this is the first thing they should read, and in some cases the authors (sometimes at the prompting of journals) actually write it with this in mind. The original function of abstracts, though, was to provide readers who didn't have the whole article in front of them with an idea of what was done and what was learned. It's typically very condensed, and therefore not the easiest thing to understand for newcomers to a field.

Also, don't start with the *Introduction*. This sounds like exceptionally bad advice, because the word "introduction" makes it sound like a great way to get *introduced* to some research. The function of the *Introduction* is more complicated than that, though. It's not designed to ease the casual or inexperienced reader into best understanding the research. An *Introduction* is often an argument designed to convince editors and reviewers that the article is worth being published. It does this by putting the research into the context of other research that the inexperienced reader hasn't encountered and may not understand. I therefore don't recommend starting your reading with the *Abstract* or *Introduction*. In fact, my next hint about reading papers is essentially that *you shouldn't start by **reading** at all.* Instead....

Hint Number Two: Look at the pictures. It sounds as if I'm sending you back to kindergarten, doesn't it? In a sense, I am. Ever since you learned to read, more of your schooling has focused on reading than on looking at pictures. Nevertheless, there is a lot to be gained from looking, with interest and curiosity, at pictures. And it's one of the easiest ways to get into a scientific article, exposing you to methods, results, and ideas. And, hopefully, the pictures will raise important questions in your mind that will drive you into the text of the article in search of answers. This works for many scientific papers because *figures* (the official name for pictures and graphs) and *tables* (which aren't really pictures, but which serve similar purposes for both the author and the reader) have always been an important part of communicating science.[1] I classify figures and tables as follows:

Explanatory figures: These are meant to explain things that are better explained in pictures rather than words. Sometimes these are diagrams of new scientific apparatus, or structures of molecules being studied or synthesized, or different mechanisms for a chemical reaction, or a biological signaling pathway.

Raw data: This is data presented just as it was collected or generated. In chemistry, it may be nuclear magnetic resonance (NMR) or infrared (IR) spectra. In physics, it may be the output of a computer model that predicts the density of states in a solid, or a photograph of stars in a distant galaxy. In biology, it may be a picture of an electrophoresis gel or a Western blot. It could be a table, too, but you have to be careful here—often the data in tables is *interpreted data*, our next classification. The reason why *raw data* isn't often presented in table form in modern scientific papers is that it can take up too much space. Most research involves doing experiments over and over, and you really don't have room to show all of that data. You can present the averages and standard deviations of all those experiments, but then you're already into the next category.

Interpreted data: This is data after it has been processed or aggregated. A chemist has run several kinetics experiments measuring different rates at different conditions. Is the relationship between initial rates linear with the concentration of reactant A, or quadratic? A graph showing the rates on the y-axis and different concentrations on the x-axis might be presented, with each experiment being represented by a point on the graph. A fitted line might show that the relationship is linear, not quadratic. That's what I call interpreted data—it's not just the direct output from one experiment but a graphical summary of the data from many experiments.

Interpreted data might also be a table of numbers resulting from calculations on raw data. An example of this kind of table, from an article I coauthored, is shown as table 7.1. This shows equilibrium constants for the formation of new Co(I) and Ni(I) complexes containing CO_2. The raw data were cyclic voltammograms from an electrochemical technique. I extracted potential shifts from this raw data, and when I plugged the potential shifts into a theoretical equation, I obtained the equilibrium constants K_{CO_2}. All of the cyclic voltammograms and potential shifts aren't presented—just the final equilibrium constants that people would be most interested in.

This table might look intimidating, but it's the sort of table someone who has taken first-year chemistry ought to be able to puzzle out. Although the names of the complexes look confusing, the bold "ligand" numbers on the table refer to another explanatory figure that shows what the non-metal parts of the complexes looked like. The reduction potentials, $E^{\circ\prime}$ and the equilibrium constants K_{CO_2}, relate to first-year chemistry topics. So, the overall message of this table—that the reduction potentials of the complexes affect how strongly CO_2 binds to these various metal complexes—is something a lot of undergraduate science majors ought to be able to figure out.

Classifying the figures and tables into categories doesn't tell you much, but the process of trying to classify them will help you think about how much you know about the techniques and concepts the paper is concerned with. If you've already taken organic chemistry, you will recognize that series of tall vertical peaks as a proton NMR spectrum, so you can easily decide that it's raw data. Good! Already you can start wondering what molecule they are taking the spectrum of, and why. On the other hand, if you get some weird closed loop shapes like the ones in figure 7.1, you may wonder whether it's the output of an experiment or interpreted data. Go ahead and read the figure caption (that's allowed!) and find out that

Table 7.1. Binding Constants, K_{CO_2}, of CO_2 to Co and Ni Tetrazamacrocycles in $(CH_3)_2SO$/0.1 M $TBA(ClO_4)$. Reprinted with permission from Schmidt, M. H., Miskelly, G. M., & Lewis, N. S. Effects of redox potential, steric configuration, solvent, and alkali metal cations on the binding of carbon dioxide to cobalt(I) and nickel(I) macrocycles. *J. Am. Chem. Soc.* **112**, 3420–3426. Copyright 1990 American Chemical Society.

| | | $E^{\circ\prime}$ | |
Complex	Ligand	$(M^{II/I})^a$	K_{CO_2}, M^{-1}
$[Co(Me_4[14]1,3,8,10\text{-tetraene})]^+$	1	−0.76	$<4^b$
$[Co(Me_2[14]1,3\text{-diene})]^+$	2	−1.31	$<4^b$
$[Co(Me_6[14]1,4,8,11\text{-tetraene})]^+$	3	−1.42	<4
$[Ni(Me_6[14]ane)]^+$	8	−1.69	<4
$[Co(Me_8[14]4,11\text{-diene})]^+$	4	−1.70	7 ± 5
meso-$[Co(Me_6[14]4,11\text{-diene})]^+$	5	−1.74	$(2.6 \pm 0.5) \times 10^2$
d,l-$[Co(Me_6[14]4,11\text{-diene})]^+$	5	−1.74	$(3.0 \pm 0.7) \times 10^4$
$[Co(Me_4[14]1,8\text{-diene})]^+$	6	−1.80	$(1.0 \pm 0.3) \times 10^5$
$[Ni([14]ane)]^+$	9	−1.89	irreversible
$[Co(Me_2[14]1\text{-ene})]^+$	7	−2.00	irreversible

[a] Measured vs ferricinium/ferrocene internal reference.
[b] An irreversible reaction with CO_2 occurs upon reduction to Co^0.

it's a cyclic voltammogram. Okay, maybe you don't know what that is—but that gives you something to find out, either by reading the text or looking up cyclic voltammetry online or in a textbook. You might guess that it's raw data just by the way it's described in the caption.

Curiosity is really the primary goal of the "look at the pictures" step; if you can get further than that, good, but just starting to *wonder* about the pictures will make getting further into the paper easier. You shouldn't stop with just classifying the figures and tables, though, especially since, in many cases, you may not have the necessary background to make a definitive classification. One thing you can do is to start to relate the figures and tables to each other. How does the data presented in this figure relate to the data in the next? How does the explanatory figure relate to the data? In some cases, you will see that the same technique in one figure is used in a second figure but is used on a different thing—a different molecule, a different material, a different set of concentrations or

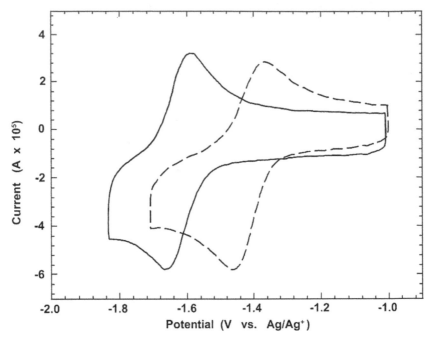

Figure 7.1. Cyclic voltammetry of $[Co(1)]^{2+/+}$ in $(CH_3)_2SO/0.1$ M TBA(ClO$_4$): solid line under N_2 atmosphere and dashed line $[CO_2] = 0.132$ M. Potential measured vs Ag/Ag$^+$ reference, approximately –0.14 V vs. ferricinium/ferrocene. Reprinted with permission from Schmidt, M. H., Miskelly, G. M., & Lewis, N.S. Effects of redox potential, steric configuration, solvent, and alkali metal cations on the binding of carbon dioxide to cobalt(I) and nickel(I) macrocycles. *J. Am. Chem. Soc.* **112**, 3420–3426. Copyright 1990 American Chemical Society.

temperatures. In other cases, different techniques will be applied to the same thing to yield the different figures. Seeing what the figures and tables have in common, and what is different among them, can help you guess at what the main point of the research is.

Hint Number Three: Decide what you want to get out of the paper. Lots of scientists have file drawers full of photocopied papers that they've only read parts of. This is changing, of course—lots of scientists are now accumulating hundreds of pdf files on their computer that they've only read parts of. But the point is that you aren't obligated to read the whole thing. If you're interested in the methods they used, focus on the *Methods*. If you're interested in the results they got, focus on the *Results*. If you're interested in what those results mean, focus on the *Discussion*. The paper may well prove interesting enough to you that you are lured into reading the whole thing, but if it's not, at least you will have gotten what you wanted out of it.

Hint Number Four: Be prepared to look up references. Because of the drive to compress as much information as possible into as few words as possible, many details and much of the background about theories and techniques aren't going to be in the paper. Maybe what you want to learn from the paper is how to synthesize pentamminerurhenium (II) chloride, but maybe all the paper says is that "Pentammineruthenium (II) chloride was synthesized according to the method of Vogt *et al.*[4]" This means you leave off reading the paper in your hand (or on your computer screen) and go find that article by Vogt *et alia*. If you want to understand the theory being used to interpret the results, you'd better find the reference that explains the theory. If the authors briefly mention a technique that you have never heard of, but there's a reference near where they're mentioning that technique, you might be able to find out more about the technique from that reference.

Hint Number Five: Learn something you didn't expect to learn. In chapter 6 we discussed how to predict what you'll learn from reading a paper. It's a useful skill that will keep you from retrieving lots of papers that end up being of no interest to you. But if you could accurately predict everything you'd find in a given paper, there wouldn't be any point in reading papers, would there? So, if you have gotten what you want out of a paper, as suggested by Hint Number Three, then move on to the sections you *didn't* read yet and get some other ideas from the paper. I guarantee that you *will* learn something new and unexpected. It may even lead to a research idea that you would never have thought of otherwise.

Hint Number Six: Figure out what the arguments of the paper are. This is not really a hint; this is what you do when you've already done all of the above, and now you want a good, comprehensive understanding of what the paper contributes to the scientific literature. If the paper is an important one in your field, and other people (like your research advisor or boss) expect you to have really read it and understood it, you should figure out the arguments. It's a big enough and difficult enough task that I have devoted the whole next section to it.

7.3 READING FOR ARGUMENTS

At some point, it will be necessary to go beyond the "First Look" hints I gave you above. Scientific papers are, after all, an extension of the "natural philosophy" that started before Aristotle, and *arguments* are the stuff out of which philosophy is made. I am not using the word *argument* here to describe a heated exchange between people of opposing points of view. I'm using it in the philosophical sense, as a conclusion that results from a set of premises. The two different meanings of the word are related to each other, of course— it's hard to have an argument without an argument!

Students in the twenty-first century have a lot of catching up to do—hundreds of years of science have to be absorbed before you're really ready to make major contributions of your own. This means that, given limitations of time, the science of the past is often presented to you as *answers* rather than *arguments*. "Atoms," for example, now seems simply to be the *answer* to the *question* of what matter is made of. But at the time that John Dalton was writing (1803) it was just an *argument*—a conclusion that he supported with experimental data. In philosophical language, the experimental data Dalton used were the *premises* upon which his conclusion was based.

Back in the first chapter I introduced the difference between *deductive* and *inductive* reasoning. But it's worth looking at inductive and deductive arguments in a little more detail. In a deductive argument, the conclusion follows *necessarily* from the premises, and adds no information that is not already in the premises. So, for example,

Premise 1: All elements can be found on the periodic table.
Premise 2: Gold is an element.
Conclusion: Gold can be found on the periodic table.

Much of science, however, rests on inductive arguments, where the conclusion goes beyond what is known from the premises. We could take another example from the periodic table:

Premise 1: Elemental beryllium conducts electricity.
Premise 2: Elemental magnesium conducts electricity.

Premise 3: Elemental calcium conducts electricity.
Premise 4: Elemental strontium conducts electricity.
Conclusion: All elements of the second column of the periodic table conduct electricity.

With inductive arguments, there is always the possibility that the conclusion *won't* hold up to further experimentation. Will the element with 120 protons, which ought to land below Ra on the periodic table, also conduct electricity? We won't know until it's created and studied.

When we were discussing the philosophy of science in chapter 1, we saw that chemists of the early twentieth century had used similar inductive reasoning to arrive at the conclusion "Elements in the last column don't react to form compounds." But in 1962, a number of Xe compounds were reported. The conclusion had to be modified to, "The elements in the last column *tend* not to form compounds, but the elements lower down in the column can be made to form compounds in some cases."

You may think that last quotation is directly from a paper in 1962. Well, it's not. If you look at the communications from 1962 where the first compounds of Xe were reported, none of them say anything general about the inert or noble gases. Perhaps, in the case of the communications from 1962 about the Xe compounds, the stunning implications of the new experimental results were so *obvious* that the authors didn't even feel the need to mention them. Generalizations about the reactivity of the noble gases didn't appear in print until later. This illustrates a very important point: *important scientific arguments are rarely contained within a single scientific paper*. Even when the papers or communications present the most important results for exciting new arguments, they may not mention these important arguments or their conclusions!

Faced with the fact that important scientific arguments might only develop through *many* different individual papers, we must once again abandon the idea of reading a scientific paper the way we would read a book. A book will usually contain some big key arguments, and the entire text will be used to support those arguments. The book will be more or less self-contained. But, just as we had to be ready to look up references in understanding the methods or background of a scientific paper, so must we be ready to read beyond the paper to really understand the *larger arguments* that the paper is concerned with.

7.4 LOCAL ARGUMENTS AND LARGER ARGUMENTS

When I have students read scientific papers for arguments, I have them find two different kinds of arguments: the *local arguments* and the *larger arguments*. A well-written paper will have some arguments that are entirely contained within the paper—those are the *local arguments*. A lot of times they won't be that exciting, because they will just be arguments that a few more important observations or data points have been taken and are reliable. The importance of these observations or data points rests on the existence of *larger arguments* that are taking place across the literature of a discipline. In the *Introduction* and *Discussion* of the paper, the authors will often refer to those larger arguments in order to explain why the local argument is important. The conclusion of the local argument essentially becomes one premise in the larger arguments presented in the *Introduction*.

Rather than jump into a real paper, with a lot of subtle and complicated science, let's make up a hypothetical scientific literature about something simple. Let's say there is a discussion amongst students as to whether the on-campus cafeteria over-salts its food. Some students try to make the argument that, yes, the cafeteria over-salts all of its food. How do they establish this? Well, one student may start by publishing a paper that shows that the black bean soup is too salty. A good paper on this topic would have to demonstrate that a proper method for determining the saltiness of the soup was employed, and that the results showed the black bean soup was in fact rather salty. The *Discussion* of the paper would have to compare the saltiness of the bean soup with a well-accepted standard for the saltiness of black bean soup, and conclude that the soup is too salty. The *local argument* of this paper might then be as follows:

Premise 1: The saltiness of a soup can be established by measuring the conductivity of the soup.

Premise 2: The proper measurement of conductivity was performed in this study to establish the saltiness of the black bean soup in the cafeteria.

Premise 3: This study showed the black bean soup to have a conductivity equivalent to a 1000 ppm salt solution.

Premise 4: Black bean soups at other dining establishments had a conductivity equivalent to a 500 to 600 ppm sodium chloride solution, and recipes in well-accepted cookbooks also yield soups of this range of conductivities.

Conclusion: The cafeteria over-salts its black bean soup.

This local argument might be contained in the *Abstract*, but it might not be fully reported there, especially in some disciplines and journals. (More about that in section 13.10.) A more reliable strategy for finding the local argument is to look at the end of the *Discussion*, or a separate *Conclusions* section, for the conclusion to the argument. You can then go back to the rest of the *Discussion* and results to see what have been used as premises to justify that conclusion.

In the *Introduction* to the paper, the author might have mentioned that other students have been concerned about the saltiness of the food in the cafeteria, signaling to the reader that the data in the paper is meant to be used in a *larger argument*, not yet completed, that the cafeteria over-salts its food. If you're trying to find out larger arguments in a paper, it's a good idea to go back and read the *Introduction* that I told you to skip over back in Hint Number One in the previous section!

Another paper, perhaps by a different author, might also refer to the discussion of the saltiness of food that has been ongoing, and may also cite, probably in the *Introduction*, the first paper on black bean soup as an example of evidence that the cafeteria food is too salty. The paper might then go on to provide a good local argument that the vegetable soup is also too salty. The larger argument, that the cafeteria over-salts its food, while still not complete, would begin to look more robust.

After quite a few papers have been published, it might be possible to construct the following larger argument:

Premise 1: The cafeteria over-salts its black bean soup, as established by:
> Nasser, M. Conductometric analysis of the saltiness of black bean soup at the Beach University cafeteria. *J. Soup Chem.* **45**, 642–651 (2013).

Premise 2: The cafeteria over-salts its vegetable soup, as established by:
> Gomez, P. & Han, D. K. Determination of sodium in a vegetable soup matrix by atomic emission spectroscopy. *J. Comest. Res.* **110**, 24–28 (2014).

Premise 3: The French fries at the cafeteria are about as salty as the usual standard for French fries, as established by:
> Mwangi, K. L. Analysis of sodium chloride residues on fried *Solanum tuberosum* fragments. *J. Fried Foods* **112**, 76–80 (2015).

Premise 4: The steamed broccoli appears not to have been salted at all, as presented in:
> Salonen, P. & Singh, R. Aqueous extraction and quantification of sodium chloride from cooked *Brassica*. *J. Cruc. Veg.* **78**, 1123–1125 (2017).

Conclusion: The cafeteria seems to over-salt its soups, but solid foods, at least of a vegetable nature, are not over-salted. More research is needed to better establish the extent of salting over a wide range of food types, especially meats, which have not yet been investigated.

The final conclusion isn't too satisfying, is it? It appears that the original hypothesis, that the cafeteria over-salts its food, may have overemphasized the importance of the soups. But we'll have to wait and see what other studies show. It is often the case that the larger arguments of science take a long, long time to really be solidified to the point where no future research is needed.

It is well worth wondering where the conclusion to a larger argument like the one above might be found. It may be found in a review paper, but it need not be. But chances are there's just too little literature in this particular area so far to warrant a review paper—anybody can read the four papers already published on the subject. Once there are fifty or a hundred papers, a review paper might be in order—especially if several papers disagree about the cream of cauliflower soup—and someone needs to come along with a critical eye and figure out which papers to trust on that soup. But you don't need a review paper to get a sense of the current state of a larger argument; you just have to read the *Introduction* on the latest paper in the field to get a sense of where the field is going. If there is ever a textbook published about the eating options on campus, the larger argument so far may be summarized in the chapter on the cafeteria.

7.5 THINKING BEYOND THE PAPER

Ultimately, when you become an experienced scientist, just determining the arguments in a paper won't be enough; you'll also have to evaluate the arguments. You'll have to start by asking critical questions like the following: Does the conclusion necessarily follow from the premises, or are there unstated assumptions that may not be true? Is the experimental evidence reliable, or have the authors failed to control all the important variables? Are there other conclusions that fit equally well with the evidence presented? Some people naively assume that, since scientific journals are peer reviewed, experts have already considered all the possible flaws and that you can take a paper's conclusions as reliable. In the next chapter, however, I will explain that peer review, as a flawed human process, can't guarantee very much.

You should also think about whether you should build on this research. Are there other experiments you can do to either confirm or challenge the author's conclusions? Are there other systems or phenomena that might be explained by the author's model or hypothesis? Would it be worthwhile to show that the model or hypothesis is more generally applicable? Do the conclusions raise yet more questions?

I can't give you a simple method for learning how to ask and answer these questions. Tracing through the literature by looking at cited references and trying to imagine how the idea for newer research might have been inspired by older research is one approach. A more effective approach might be to join a research group where others are tossing ideas about in conversation—you will find that many ideas are generated, thought about carefully, and then either pursued or rejected. Many more are rejected than are pursued! We'll discuss this further in chapter 9.

FOR FURTHER STUDY AND DISCUSSION

1. Find an experimental scientific article in the literature from some years back. Older scientific articles are more likely to be understood by undergraduate students because the science being done then is often the science you learn about in your textbooks today. For second- or third-year chemistry students, I recommend articles from the 1950s and 1960s. Look at all the figures and tables and try to classify each one as being an explanatory figure, raw data, or interpreted data. What can you learn from just looking at the figures? What questions about the study arise after you look at the figures?
2. Your instructor will assign you a good scientific article in your field that has both local and larger arguments. Try to identify the local arguments and the larger arguments, and the premises that support the conclusion. The premises that support the conclusion of the local argument may be too technically advanced to understand fully at this point in your education, but you should be able to get some idea of what experimental results are being used.

ADDITIONAL READING

Pain, E. How to (seriously) read a scientific paper. *Science* (2016). doi:10.1126/science.caredit.a1600047.
Ruben, A. How to read a scientific paper. *Science* (2016). doi:10.1126/science.caredit.a1600012.

These two online *Science* articles give some other, different perspectives on how to read scientific papers. I think the suggestions are geared more towards graduate students and professionals, however.

CHAPTER EIGHT

PEER REVIEW

LEARNING OBJECTIVES

After reading this chapter, you will be able to:

- identify advantages and limitations of peer review
- explain how the current system of peer review is different from past systems for deciding what gets published
- explain the central compromise of peer review—a compromise between a push for greater quality control balanced with the need to be cost-effective, prompt, and open to new ideas
- identify the main steps in the peer-review process
- identify attempts and suggestions that have been made to improve peer review

8.1 BENEFITS AND LIMITATIONS

Just about all scientific literature published today, both journals and books, is peer reviewed. Peer review is an important part of science, and of scholarship in other disciplines as well. In chapter 2, we looked at how it was the most obvious example of the *organized skepticism* that Robert K. Merton had as one of the norms of his *ethos* of science. In chapter 13, when I discuss the assembling and writing of a scientific paper, I will constantly remind you that you are writing not just for readers but also for reviewers, fellow scientists who will be skeptical and questioning about every table, graph, and sentence. In this chapter, we'll look more closely at the peer-review process, examining its advantages and disadvantages.

The progress of science depends on the work of many scientists, working independently. Scientific conclusions are often assembled from, and strengthened by, multiple papers by multiple authors. No one scientist can do all the experiments, and do all the thinking, necessary to build up a modern scientific conclusion from nothing. Scientists rely on the quality of the papers they read when planning experiments, interpreting those experiments, and assembling complex scientific arguments. One of the reasons for peer review is to provide some level of quality control for published work, so that the experiments, observations, and conclusions of that work can be trusted.

I deliberately wrote *"some level* of quality control" because there are practical limits to how much peer review can do to ensure quality and still have an active, productive scientific literature. For example, if an author reports the results of a faulty experiment, or deliberately lies about the results of an experiment, peer reviewers may not be able to detect the problem, because reviewers typically do not have the time or resources to replicate the experiments reported by authors. If the result is important, somebody may eventually redo the experiment as the first step in doing further research in the area, and problems of faulty experiments or dishonesty may then be uncovered—but that is typically only after the paper has already been published. Problems in the interpretation of results are more likely to be caught by peer review, but if reviewers accept the same faulty premises as the authors, or have logical skills no better than those of the authors, such problems can still go undetected.

In principle, you could solve these problems by having more reviewers, demanding that reviewers repeat all of the experiments and get the same results, and requiring all these reviewers to agree completely on the interpretation. With the complexity and cost of doing modern research, however, this isn't a workable solution.

Even if the practical obstacles were removed, there would still be difficulties; if one of the reviewers wasn't a good experimentalist and couldn't repeat the results, or if one of the reviewers couldn't grasp the new idea that the author was trying to present, an article describing good science might never make it into print.

Peer review, then, is always a bit of a balancing act. If a journal's system for peer review is too loose, lots of bad articles will get published, and people won't want to read the journal anymore. If it's too stringent, very few articles will make it into print, and the ones that do will probably be just the safest, least controversial, and least innovative articles. There has to be a compromise; inevitably some worthy articles are rejected, and some faulty articles are published.

Another limitation of peer review as a system of quality control is that not all of the reviewers are the ethical scientists they should be. Reviewers typically work in the same field and specialty as the work being submitted and could be competitors who don't want other researchers' work or ideas to succeed. In such a situation, they might exaggerate minor flaws in papers they review, hoping to delay or thwart publication of their competitors' work. On the other hand, some reviewers might want to promote the work of a scientist they know personally and ignore some of the problems with the work.

Some people look at all these problems with peer review and wonder whether it is worth the time, cost, and effort it takes. Well, it does provide *some level* of quality control. Papers that have no worthwhile results or obviously flawed interpretations of results can be rejected. For the papers that are accepted for publication, having more people looking at a paper before it's published has a very good chance of making the paper better. Science has always worked best as a communal enterprise, because different people see different things when they look at theories, experiments, data, and interpretations. It's a good idea to have at least some of these alternative views brought out for consideration *before* publication. In some cases, the alternative views are incorporated in the manuscript; in other cases, they are rejected, but the article is still published; in still others, the alternative views win over the editors and the paper is rejected. The important thing to remember about peer review is that it doesn't guarantee any specific level of quality. It is just a way of trying to choose the best work to publish in journals, and to make the work that is published in journals better.

Peer review was not consciously designed at one point in history. Like scientific journals, it has evolved over many centuries, and the diversity of peer review practices echoes the diversity of journals. The challenge for organizations and journals over the last few centuries has always been to balance the desire for stringent quality control with a philosophy of free publication that allows for maximum exchange of information and ideas. This balance is one that modern journals still wrestle with as the number of scientists seeking to publish continues to expand.

8.2 HISTORICAL BACKGROUND

England: The early history of British science tended to lean a little more to the side of freedom than control. The English philosopher Francis Bacon's *Novum Organum* in the seventeenth century was a call to open up scientific investigation to new ideas instead of relying on the authority of the experts of the ancient world or the church. Inspired by Baconian philosophy, the Royal Society in London took as its motto *"Nullius in verba,"* Latin that can be roughly translated as "Take nobody's word for it," a direct challenge to the old authorities. (For more on the Royal Society, see chapter 4.) The early meetings of the Royal Society often included demonstrations and experiments, so the Society could to some extent pretend they were "taking nobody's word for it."

However, as soon as Henry Oldenburg began publishing the *Philosophical Transactions of the Royal Society* in 1666, scientists *were* putting trust in other scientists' words, and the matter of the trustworthiness of published words became an issue. Oldenburg relied on a network of personal acquaintances to establish the trustworthiness of authors. Articles that made it into the *Transactions* were either by fellows of the Royal Society (whose membership was by election and invitation only) or by people whom the members knew personally—such as the network of correspondents throughout

Europe that Oldenburg had cultivated even before the founding of the Royal Society. Oldenburg's successors, in the position of secretary of the Royal Society, pretty much had total control over what showed up in the *Philosophical Transactions*, and membership in the Society or networks of personal acquaintance determined what was published. The main problem with this arrangement was that some people were elected to the Royal Society for mostly political reasons, not for their scientific training or ability, and the quality and trustworthiness of contributions from some of these fellows wasn't particularly good.

In the mid-eighteenth century, Sir John Hill, a botanist, provided an incentive for the Royal Society to change this publication policy. Sir John Hill was a botanist who was never elected to the Royal Society. Sir John's difficulties in getting elected to the Royal Society weren't because he couldn't make valuable scientific contributions but because he wasn't good at being tactful in his criticisms of others. People didn't like him! He resented seeing fellows of little or no scientific training publishing articles of dubious merit in the *Philosophical Transactions*. Around 1750, he started publicly ridiculing the Royal Society and its *Transactions* with broad parodies of some of the more unscientific research found in the *Transactions*. Such ridicule led the Royal Society to take action to improve the quality of the *Transactions*; control of the publication was no longer left to the secretaries but was given over to the Society as a whole. The Society formed a Committee of Papers that would decide which papers deserved publication. (Although these efforts are sometimes characterized as the beginnings of peer review in science, it should be noted that the Royal Society of Edinburgh's *Medical Essays and Observations*, established in 1731, already had something of a system of peer review.)

The Committee of Papers served for several decades to decide which papers were published in the *Philosophical Transactions* until another wave of critics, this time within the Society, urged further reforms. Because the committee was not made up of experts across all fields, it was sometimes the case that the committee would vote for or against the publication of a manuscript even though no one on the committee had any expertise in the subject matter of the manuscript. This struck some critics within the Society as a poor way to make objective decisions. It gave too much power to a few fellows of the Royal Society who could reward friends in high places with publication, while rejecting good papers by foreigners and others who weren't well known to those on the committee.

In the 1830s, the system was again reformed, largely following the recommendations of William Whewell, whom we first met in chapter 1. Whewell suggested that no paper should be published in the *Philosophical Transactions* until after other distinguished scientists, working in the same field as the paper, had written reports on the paper. Whewell intended these reports to be published on their own in a new publication, the *Proceedings of the Royal Society*, to further publicize the work being published by the Royal Society. The publication of these reports in the *Proceedings* did not go on for long. The written reports gradually came to be more like the unpublished reviews of referees that are a part of modern peer review, and these were used to determine what was to be published and what was not. The evolution of publication of the Royal Society, then, was in the direction of more expert control of publication between the seventeenth and the nineteenth centuries.

France: Whewell's proposals for publication reform, which were in the direction of more quality control, were inspired by policies from the Royal Academy of Sciences in Paris. From the very beginning, the politics of the Paris Academy of Sciences was different from that of the Royal Society of London. The Academy was founded in 1666, about the same time as the Royal Society. Despite the fact that both institutions had "Royal" in their names, the French crown had a great deal more control over the Paris Academy than any English royalty ever had over the Royal Society. Election to the Paris Academy was subject to approval of the crown, and the Academy was expected to provide the crown with expert advice on granting privileges for inventions (somewhat analogous to our modern patent system) and any other matter requiring technical expertise. These duties made them what we might now call government scientists.

The Academy had its own annual publications, the *Histoire* and *Mémoires*, that published members' contributions. In 1750, the Academy initiated a new journal for publishing the work of people outside the society, commonly known as the *Mémoires de Savants Étrangers* (its official title was much longer). The Academy's influence extended beyond their own publications, however; in the 1700s, the Academy also had a lot of control over what appeared in the *Journal des Sçavans*, France's first scientific journal, and other publications. This is because, at this time, the French government still had a system of government censorship of all printed materials—almost everything that was printed had to be approved by the Royal Board of Censors. Anything approved by the Paris Academy of Sciences, acting as the representatives of the crown, was essentially pre-approved by the government and didn't have to go to the Board of Censors. The result was that the Academy could effectively play the role of reviewers or referees even for work not directly published by the Academy.

During the eighteenth century, control over the Academy's publications was exerted by a powerful *Comité de Librairies* that reviewed all papers presented to the Academy during their meetings. Papers were sent out by this committee for review by other members of the Academy, who viewed the claims of the papers skeptically, made recommendations for further experiments, and edited the papers for length and style; in some cases, the editing could even be considered to have changed the scientific content of the papers themselves. When possible, the experiments described in the papers were repeated by the people assigned to report on the submissions. The Paris Academy's procedures were another predecessor to modern peer review—but it could be argued that, especially in the case of the *Mémoires de Savants Étrangers*, it was more review by an elect few, as membership in the Paris Academy was always restricted to a small number of scientists.

After the French Revolution of 1789, there was no monarchy—and, for a brief time in the 1790s, no Academy, either. The new, more democratic government that emerged in 1795, however, set up a new institute incorporating all branches of scholarship, and the First Class of this institute more or less corresponded to the old Academy of Sciences. Publications of the *Mémoires* and the *Histoire* resumed, as well as a renamed version of the *Mémoires de Savants Étrangers*. The practice of sending papers presented to the Academy to members for detailed reviews or *reports*, sometimes involving replication of experiments, was also resumed. These were the practices that Whewell and Herschel of the Royal Society so admired, and tried to imitate, in the Royal Society in the 1830s.

But even as the Royal Society was trying to imitate the Academy, the Academy was starting to find their system unworkable. There were too many submissions sent out to the limited number of Academy members. Full reports on submitted papers became more and more rare, and the publication of the *Mémoires* was slow. A demand for more up-to-date reporting of scientific results prompted the Academy to start a new journal in 1835, the *Comptes rendus*, which allowed prompter publication of shorter papers. The *Comptes rendus* allowed papers presented at the regular Monday meeting of the Academy to be in print by the following Saturday. These papers did not undergo any rigorous process of review; members of the Academy who voiced doubts or criticisms about the presented works at the Monday meeting could have their comments published alongside the paper, but they had to provide written summaries of their comments to the secretary by the end of the meeting at which the paper was originally presented.

For many scientists, however, getting published in the *Comptes rendus* actually involved an additional review process. This is because people who were not Academy members had to have an Academy member introduce or sponsor their paper before the Academy, and the member's name was attached to the paper as the presenter. Members would not present other scientists' work if it might tarnish their own reputation, so they effectively acted as referees in deciding whether a paper was important enough and good enough to be presented at the Academy.[1] In the Paris Academy, then, the trajectory was almost the opposite of what happened in the Royal Society of London; it went from strict control by appointed experts to a more open system that relied more on personal acquaintance.

Germany and Elsewhere: In journals published outside these two prominent national academies, the responsibility for assuring the quality of published works largely rested with their editors. Many nineteenth-century journals were closely linked with their editors, who, typically, were well-known experts in their fields. People tended to trust the editors' judgment as to what was good scientific work and what wasn't. Some of the oldest chemistry literature that is still referenced today appeared in *Annalen der Chemie und Pharmacie*, edited by Justus von Liebig, an important German chemist of the mid-nineteenth century. People knew the journal as *"Liebig's Annalen,"* and they trusted the contents of the journal because they trusted Liebig. After his death in 1873 the journal was renamed *Justus Liebigs Annalen der Chemie und Pharmacie* in his honor. Likewise, *Annalen der Physik*, the journal in which Einstein would publish his work in the twentieth century, was a descendant of Gilbert's *Annalen der Physik*, which Ludwig Wilhelm Gilbert edited for twenty-five years. This journal would become known as *"Poggendorff's Annalen"* under the fifty-two-year editorship of Johann Christian Poggendorff. The tradition of relying on journal editors to make decisions about submitted manuscripts without formal peer review continued into the twentieth century in Germany. In 1936, Albert Einstein was reportedly surprised and angered when an article he sent to the U.S. journal *Physical Review* was sent out for peer review. He apparently thought that the editor should just publish it or reject it, as strong editors had done in German journals for more than a century. *Nature*, one of the world's premier scientific journals, published in the United Kingdom, did not have a formal requirement for peer review until 1973, although the editor did have an informal system of handing out papers to scientific acquaintances for their opinions starting around the middle of the twentieth century.

Modern Peer Review: Modern peer review is really a descendant of all these traditions; different societies and journals have tried different things and borrowed ideas from one another. The idea of written reports or reviews on submitted manuscripts has come down to us from the Paris Academy, somewhat altered through its adoption and evolution by the Royal Society. Membership in national academies or societies still grants some privileges, but these are declining as journals try to be more open, transparent, and fair in deciding what gets published. Most journals still have editors (and associate editors) who have the final word on what gets published and how, but they tend to go along with peer reviewers rather than personally taking responsibility for the quality of all submissions.

One of the biggest differences between historical models for review and modern review are the number of people involved in reviewing; if you send a paper to a journal, you are likely to get at least one back to review. This makes modern review much more a system of *peer* review than the *expert* review that tended to prevail in the Paris Academy of Sciences, the Royal Society, or the German journals headed up by very strong editors. As we saw with the Paris Academy, relying on a small number of experts results in overwork of the experts. In today's world, the problem might be even greater, with so many more people trying to publish work. The broader participation of scientists in *peer* review has both advantages and disadvantages over *expert* review; science is far less likely to be controlled by just a few people with lots of power, but manuscripts may not be reviewed by the people who can make the most knowledgeable judgment on the significance or quality of the work.

8.3 MODERN PEER REVIEW IN PRACTICE

Peer review does vary from academic discipline to academic discipline, and from journal to journal, so it is impossible to give a complete description of exactly what happens in all cases. But a rough outline of how it works is as follows:

1. *An author submits a manuscript.* Not so long ago, manuscripts were mailed to the editor of a journal, with a polite cover letter addressed to the editor. Within the last twenty years, this has changed so that, in many cases, the manuscript is uploaded via the World Wide Web (WWW), and the information that used to be provided in letter form is now just part of an online form. That way the journal can ensure that they will get all the information they need.
2. *An editor takes a first look at the manuscript.* An editor will sometimes make an initial judgment that the paper just doesn't fit into the journal's scope and reject it. The editor may also reject it if it's obviously not well enough prepared to undergo review

by reviewers. If the paper isn't rejected at this early stage, the editor will often hand it off to an associate editor. Except in very small journals, there are almost always a number of associate editors with certain specialties. Justus von Liebig could make some kind of judgment about almost any chemistry manuscript submitted to him in the mid-nineteenth century, but now even those journals specific to a sub-discipline will have associate editors for particular sub-sub-disciplines, and the journals may have a large number of associate editors just to handle the large number of submissions.

3. *An editor or associate editor assigns the paper to reviewers.* How this is done varies, and the number of reviewers can also vary, although it is usually two, and sometimes three. One important criterion for reviewers is that they do not have an obvious conflict of interest in reviewing the manuscript; the reviewer shouldn't be a collaborator of the author or work at the same institution. Ideally, the associate editor knows the best experts to judge a particular paper and will send it to them. But too much reliance on the very best experts raises the problems that the Paris Academy of Sciences ran into in the nineteenth century; too few top experts being asked to review too much work, much of which might not be worth their time. Often, journal editors will tend to send review requests to people who have submitted other papers to the same journal; some people view this as "fair" because the people creating the most work for reviewers then also end up doing the most reviewing. Some journals now will even ask authors if they have suggestions for reviewers. In some ways, this makes a lot of sense; as science becomes more and more specialized, the author might well know better than the editors just who will understand the paper's content. Of course, asking authors to suggest reviewers can also be abused; the author can simply request reviewers he or she knows will be predisposed to accept the conclusions or significance of the paper. When asked for potential reviewers, I think it's a good idea to request reviewers whom you respect and are at the very top of the field. It may make it harder to get your work published, but you'll get the best (if maybe the most painful)

feedback, and your name might become more familiar to those at the top of the field.

4. *The reviewer receives the manuscript and accepts the responsibility to review.* Back when people communicated mostly by phone and "snail mail," the associate editor would send out the manuscript together with a request to review the work; reviewers were asked to send it back if they didn't think they were qualified to review it or didn't have the time. Now, potential reviewers often get an email with the title and abstract of a paper to be reviewed, asking whether they would consider reviewing the paper. Only after accepting the responsibility to review does the reviewer get a chance to see the manuscript. This not only saves the time and expense of sending out manuscripts to people who don't end up submitting reviews, but it also provides something of a check on the unethical practice of reading a manuscript just to see what other people are doing before it is published, without actually contributing a helpful review.

   In committing to review a manuscript, a reviewer is implicitly or explicitly assuming a set of ethical obligations to follow. Reviewers, for example, are often required to certify that they do not have conflicts of interest. A complete and detailed list of ethical obligations can be found online at the Committee on Publication Ethics website (publicationethics.org, search Ethical Guidelines for Peer Reviewers). It's a fairly long list, but it basically boils down to treating the authors fairly and not taking advantage of your role as reviewer. The manuscript being reviewed is confidential and is not to be shared with anyone; what you think and write about the manuscript is also confidential. You shouldn't reveal, even after publication, that you are or were a reviewer of the manuscript (unless it's some form of *unblinded or attributed* peer review; see below). You shouldn't use the information from the manuscript to get a jump on the competition.

5. *The reviewer reads the manuscript and submits a review.* The request for review will typically include instructions for the reviewers. The reviewer is usually asked to review many aspects of the manuscript. Any problems—with the experiments, the quality of data, the statistical treatment of data, or the conclusion—must

of course be pointed out, but the reviewer is also asked to comment on whether the results and conclusion are new enough or significant enough for publication in the journal. The length of the manuscript, the clarity of the writing, and the usefulness of the figures are also fair game for criticism. Even pointing out typographical errors is appropriate. A summary judgment is also typically requested: should the manuscript be published as is, published after some revision, or not published?

6. *The associate editor collects the reviews and sees if there is a consensus on the summary judgment; if there is no consensus or majority, he or she may send it for further review.*

7. *The decision and reviews are sent to the author(s).* The decision whether to publish or not publish rests with the editor or associate editor, based on the reviews submitted by the reviewer. Regardless of what the decision is, all reviews are sent to the author of the manuscript. The decision to "publish as is" is rare; usually some revision is requested. The amount of revision can range from minor editorial changes to a complete rewrite to even doing more experiments.

8. *The author responds.* Decisions can be, and are, appealed. A decision not to publish is a hard one to reverse, although I know of cases where it has happened. The author must convince the editor that the criticisms of the reviewers are wrong; that perhaps the reviewer has misunderstood the paper (in which case the paper should perhaps be clearer!) or was making unwarranted assumptions. The editor may send out the paper to yet more reviewers to see if they have different views of the paper. It is very important, in arguing against the conclusions of the reviewers, not to make things personal, but to be very polite and focus on the scientific issues at hand rather than the imagined personality or intelligence of the reviewers.

When the decision on the manuscript is to "publish with revisions," the author may acknowledge that the criticisms are valid, make the necessary changes, and resubmit the manuscript. Sometimes an author may find some of the criticisms invalid, and refuse to make some changes; once again, it is up to the author to convince the editor that the reviewers are wrong. If major revisions are made—or if the author refuses to make major revisions that have been recommended—the paper may go out to reviewers again. These may be the original reviewers or new reviewers, depending on the situation. In particularly contentious fields of science, it's easy to see how the peer review process could drag on for many months or years if both author and reviewers stand their ground.

In many cases, though, the author tries to appease the reviewers and editor by making the recommended changes. Bold claims might be removed and replaced with the more careful language we will discuss in the chapters on scientific writing style. The careful writing style scientists use has probably evolved as an adaptation to decades of battles between authors and reviewers. In modern times, it is often more important to a scientist to get *something* published than to get the full scope of his or her ideas published, so authors are inclined to accept many recommendations for revisions without putting up too much of a fight. This doesn't mean that they are always happy at the end of the process, but there is always the possibility of getting more advanced ideas into future publications.

8.4 SOME PROBLEMS WITH PEER REVIEW, AND SOME POSSIBLE SOLUTIONS

We already looked at some inherent tensions in peer review in the first section of this chapter. Beyond these, however, there also are more practical problems, and there is a lot of discussion in the major scientific journals about peer review and how to improve it. These discussions are especially intense in medical journals. Here are some of the perceived problems:

1. *Reviewer bias.* If reviewers are reviewing the manuscript of a competitor, don't they have an interest in delaying or thwarting publication? On the other hand, if the reviewers have a friendly relationship with the author(s), could they be biased in the opposite direction, being insufficiently critical? Although reviewers often have to certify that they don't have a conflict

of interest in reviewing the manuscript, some fields are small enough that there aren't any qualified reviewers who don't have at least some sort of relationship with an author.

2. *Delays in publication.* In fast-moving fields, the delays in publication caused by a lengthy peer-review process can hinder the development of the science. In medical fields, this can potentially affect the availability of new treatments or result in the continued use of treatments found to be harmful. In other cases, scientists may lose priority in receiving credit for discoveries because publication was delayed or may miss out on funding opportunities.

3. *Some reviewers get overworked.* Although editors do, in general, try to spread the burden of reviewing around to as many people as possible, some reviewers whose opinions and insight are particularly well trusted are likely to get many more manuscripts to review, and the quality of the reviews may suffer.

4. *Peer review may not improve decision-making about what gets published and may introduce bias into the scientific record.* One study, using a manuscript with eight "weaknesses" deliberately added, showed that, on average, reviewers found fewer than two of the weaknesses.[2] Other studies have attempted to uncover biased decision-making based on the institutions where authors work, the countries that they come from, or the author's gender; while direct evidence of such biases could not be found, there were patterns which suggested that such biases could exist.[3]

5. *Peer review may make it harder for new ideas to get published.* Back in chapter 1, we saw how Kuhn was concerned with *paradigm shifts*, where many of the assumptions and methods of a discipline are questioned and new methods and issues take their place. These revolutions often don't go smoothly; the new methods and issues are likely to encounter a lot of criticism in the peer review process. Reviewers who have invested a whole career in the old paradigm may be resistant to the new methods and ideas.

In order to address some of these concerns, a number of modifications have been suggested, and tried, to improve the peer-review process:

1. *Double-blind peer review.* In the natural sciences, peer review tends
   to be *single-blind*; this means that the reviewers know the names of
   the manuscript's authors, but the authors don't know the names
   of the reviewers. In other disciplines, *double-blind* peer review is
   the norm; only the editor knows the names of the authors and
   reviewers. A change toward double-blind review could, *theoreti-
   cally*, lessen the problems of bias against particular authors. But
   in many narrow fields of research it's often well-known who is
   doing what kind of research, and many reviewers probably have
   a good guess as to whose manuscript it is, especially if the authors
   cite a lot of their own previous work. For that matter, it isn't even
   too hard to guess who some of the reviewers are in single-blinded
   peer review, just by the kinds of comments they make, or the
   references they suggest the author include. While some people
   I know in other fields think that double-blind review is an obvious improvement on single-blind review, I haven't seen too much
   call for it among peer-review critics and reformers. Some studies
   of peer review in the medical literature demonstrated that double-
   blinding wasn't very successful in keeping the identity of the
   authors secret;[4] and a variety of studies have shown little effect on
   the actual quality of peer review.[5]

2. *Unblinded peer review,* also called *attributed peer review.* Under this
   system, neither the authors nor the reviewers are anonymous.
   The idea here is that reviewers will not write negative reviews
   that they can't substantiate or defend; criticisms motivated
   primarily by the desire to delay or suppress publication will
   therefore be minimized. A disadvantage of this system, however,
   is that reviewers might hold back from making necessary criticisms if they think they will be making enemies of the authors,
   who now know their names! Another potential downside is that
   scientists may be less likely to be willing to review papers if their
   names will be attached to reviews, and this was actually found
   in one study of attributed peer review.[6] The reasons for this
   might not be bad reasons: younger scientists may be reluctant
   to criticize the work of more established scientists who may be
   involved in reviewing grant proposals,[7] or scientists may just be
   reluctant to ask critical questions of authors that might turn out

to be ignorant questions.[8] Sometimes "ignorant" questions turn out to be insightful, while at other times they're just ignorant.

3. *Public peer review.* Taken to an extreme, unblinded peer review becomes public peer review, where the reviews of papers are published alongside the papers themselves. This is similar to what Whewell envisioned for the reform of publishing practices for the *Philosophical Transactions*, except that he envisioned that the reviews would be substantial enough to be published on their own in the *Proceedings of the Royal Society*. The reading public can read both the edited manuscript and critiques of the manuscript; conflicts between authors and reviewers are out in the open, rather than being resolved in one direction or another by an editor's decision.

4. *Limiting the role of reviewers.* Some publications, most notably *PLOS ONE*, only ask reviewers to review manuscripts for ethical and technical soundness, not for whether the results reported are likely to be important to a field. This speeds the acceptance process and reduces the chances of innovative papers being suppressed by subjective judgments of the reviewers about how important the work is; it even makes it possible for negative results to be published. This system is made workable for *PLOS ONE* because it isn't a printed journal; it is entirely Web-based, so there are no limitations on space.

5. *Publish before review.* Some European geoscience journals have taken up the practice of publishing a paper online for public discussion even while traditional peer review is taking place. After both public comment and peer review, a revised version is published in the print edition of the journal.

Similar to this practice are the "preprint servers" that have gained a lot of popularity in physics and mathematics and are also getting a foothold in biology and chemistry. *ArXiv* (pronounced "archive," the X is meant to be the Greek letter *chi*) was founded in the early 1990s primarily for physics papers, but soon astronomy, mathematics, computer science, and other disciplines began participating. Many of the preprints are also submitted to regular print journals as well, where they undergo regular peer review. *Nature, Science,* and many other journals now accept papers for review that have previously been submitted to a preprint server.

The founding of both *BioRxiv* (2013) and *ChemRxiv* (2016) was delayed in part because of two major concerns. One was that preprints would not be recognized as legitimate claims of discovery but only serve as a way to give information to the competition. The other was that many of the editors at traditional journals considered publication via a preprint server as "prior publication" and therefore did not accept articles deposited on the servers for peer review. Because the American Chemical Society (ACS), one of the largest publishers of chemistry journals in the world, sponsored *ChemRxiv*, authors felt more secure in depositing their preprints with the server. As of the founding of *ChemRxiv*, 20 of the 50 ACS journals had agreed to accept articles for review that had previously been submitted to a preprint server.

There never has been, and there probably never will be, a single, uniform process for peer review. Peer review will always be a balancing act, setting the objective of maintaining a certain level of quality and importance in published papers against the need for efficiency of publication and openness to new ideas and new scientists. Authors, reviewers, and editors will make errors, but we must hope that whatever system is used will bring together the collective judgment of all these participants to make the scientific literature both trustworthy and open to new ideas and information.

FOR FURTHER STUDY AND DISCUSSION

1. Imagine you were starting a new scientific journal tomorrow. Would you have a system of peer review? What would you expect the process of peer review at your journal to accomplish? What kind of peer-review process would you implement— would it be double-blind, single-blind, attributed, or public? What would you do to minimize some of the problems that have

been known to occur with peer review? What would you expect the process of peer review at your journal to accomplish?

ADDITIONAL READING

Akst, J. I hate your paper. *The Scientist* **24,** 36–41 (2010).

A good introduction to some of the problems of peer review and what people have done to try and remedy these problems.

Crosland, M. P. *Science under Control. The French Academy of Sciences, 1795–1914* (Cambridge University Press, 2002).

The books by Crosland and by Hahn (mentioned below) together give a detailed view of the workings of the Paris Academy over almost two and a half centuries, including their publication and peer-review practices.

Csiszar, Alex. Objectivities in print. *Objectivity in Science*, 145–169 (Springer, 2015). Web.

Csiszar, Alex. Peer review: troubled from the start. *Nature* **532,** 306–308 (2016). Print.

Csiszar, a historian, looks at the evolution of peer review, mostly focusing on the Royal Society of London.

Godlee, F. & Jefferson, T. *Peer Review in Health Sciences* (BMJ Books, 2003).

A good overview of attempts to make peer review more scientific by empirically studying how it works in practice.

Hahn, R. *The Anatomy of a Scientific Institution. The Paris Academy of Sciences, 1666–1803* (University of California Press, 1971).

PART THREE

PLANNING, DOCUMENTING, AND PRESENTING SCIENCE

CHAPTER NINE

STARTING RESEARCH: A DIFFERENT "WHAT SHOULD WE DO?" QUESTION

LEARNING OBJECTIVES

After reading this chapter, you will be able to:

- explain the difference between divergent and convergent thinking, and their roles in the process of coming up with research ideas
- use general categories of "I wonder" statements to generate statements that might lead to research ideas
- use critical questions to narrow the focus on a few research ideas
- use visualization and concept maps to both refine ideas and generate new ideas

9.1 THE IMPORTANCE OF CREATIVITY

Where do ideas for scientific research come from? Very often, they come in part from the work of previous scientists, as we discussed at the end of chapter 7. Scientists read or hear about the work of other scientists and ask critical questions that require further research; or they are inspired to apply the ideas of other researchers to new systems and phenomena. However, even when the connection to previous research is obvious, there is usually an element of creativity involved in moving to the next research project, especially in the most highly valued research.

William Whewell, who came up with the hypothetico-deductive theory of science we saw in chapter 1, emphasized the need for "bold invention" in coming up with hypotheses. These days, he probably would have used the term "creativity" rather than "bold invention," but the message is the same. Scientists need to be creative people, both in coming up with research ideas and in solving problems.

Creativity in coming up with research ideas isn't only limited to those instances in which coming up with a new hypothesis is needed. When research is driven by new hypotheses, we typically call it *hypothesis-driven research*. But at other times, the research idea can just be an open question about how the world is, and what is in it, and we would describe this research as *exploratory research*. We can look back to the examples of Tycho Brahe and Johannes Kepler: Brahe was doing exploratory research in the sense that he was just making more detailed measurements of the orbit of Mars than anyone had ever done, exploring where in the night sky Mars appeared over time. Kepler, on the other hand, looked at the orbital data that Brahe collected with the hypothesis that the orbit about the Sun should fit a simple mathematical function. These days, some people (especially agencies that give money for research) favor hypothesis-driven research over exploratory research, but both have been very important for the progress of science over the last 350 years.

This book was written for a course that I have taught for many years, and the course requires students to propose and do a modest research project that they could accomplish at their homes, dorms, or apartments. I'd give them a few examples off the top of my head, show them some things students had done in past years, and ask the students to be creative. I had the students get together in groups and discuss their plans and ideas. People came up with projects, but many weren't very creative; students did projects that were pretty much like projects done in previous years, or I'd have a number of students in a given semester doing very similar things. I got bored reading all the rather similar research proposals and results that were fairly predictable. I started looking into what was known about creativity, and I found that the way I was asking students to be creative wasn't actually very helpful in getting them to develop creative ideas.

There are lots of theories of creativity and lots of different strategies for trying to encourage greater creativity, but they usually have a few elements in common. First, they recognize the importance of both *divergent thinking* and *convergent thinking*. Divergent thinking is most important at the beginning of the creative process. You have to let your brain wander over all sorts of different territory, *diverging* from a straight and narrow path to find as many possible research ideas as possible. Later in the process, it is important to *converge* on a few ideas, develop them further, and explore how workable or profitable they might be.

9.2 DIVERGENT THINKING ON A BIG SCALE

Before I researched creativity, I had a sense of the importance of divergent thinking, although I didn't call it that. I asked students to come up with at least three ideas and pick the best one. That was divergent thinking on a small scale. What I have found in recent years is that three ideas aren't nearly enough! One thing I didn't realize was that when *I* was thinking up research ideas, I would only concentrate on three or four, but that many more would pass through my head and be thrown out without too much thought. As a scientist with years of experience, I had just gotten in the habit of thinking divergently and convergently in a big way, rather quickly, so I didn't really notice just how many ideas my brain was really sifting through. But if you're just starting out as a science student, you need to *make sure* you are really generating a lot of ideas, and the best way to ensure that you do that is to force yourself to *write down* a lot of ideas.

Divergent thinking is the sort of thing that people often think of as "brainstorming"—although a lot of things that people call "brainstorming" aren't really what the inventor of that term had in mind, and they aren't always optimized for developing new ideas.

Often, we hear about "brainstorming" in the context of people getting together in small groups and "tossing around ideas." But research has shown that discussing ideas in a group doesn't create the largest, or broadest, collection of ideas. If you are slow to come up with ideas on your own, you are more likely to think about the ideas other people are talking about rather than thinking of something really new or different. It is easier just to propose some variation on what other people are talking about. I therefore no longer allow my students to discuss ideas with each other until after the first research ideas assignment. If you are forced by the first assignment to come up with twenty or thirty ideas *on your own*, you will have to put more effort into making your own mind range broadly over a lot of topics and work on a lot of different ideas. It is certainly more *difficult* than discussing ideas in a group, but nobody ever said creativity was *easy*. Coming up with twenty ideas you can write down is a good mental exercise. Just as physical exercise makes your body stronger and more flexible, this sort of creativity exercise will make your brain stronger and more flexible.

One obstacle to divergent creative thinking is fear. So much of your education thus far, especially in the sciences, has been focused on getting the right answer, and you've been penalized for getting the wrong answer! It's therefore only natural that you would be afraid that your research questions might be criticized. There is a place for criticism and fear, but it is later in the process, during the convergent thinking stage. In the divergent thinking part of the process, we need to be fearless. That's another good reason not to be in a group during this initial process—you don't want to be fearful about what other people might think about your ideas. Some of you may feel plenty confident about your ideas, but often the people who don't feel confident about their ideas actually have better ideas.

One way to get around fear in the first assignment is not to come up with complete research ideas—but just to start out with questions. As you've probably heard before, there's no such thing as a dumb question. One way to make sure that these questions aren't too influenced by fears is to begin your thoughts about research ideas as statements beginning with "I wonder...." There is never anything *wrong* about wondering.

Even though I am emphasizing divergent thinking at this point, you still need to impose some restrictions. Statements like "I wonder why my boyfriend is such a jerk," or "I wonder if the Mariners will win the World Series" aren't going to lead to good research questions. It may therefore be a good idea to limit the questions somewhat—we have to inject just a little bit of constraint even at this early point. In my course, where the students' household science projects can range over many different areas, I usually ask my students to come up with four major *classes* of "I wonder" statements, which can span a very broad range of topics, keeping in mind that the ultimate goal is a science project they can conduct at home. In other situations, such as an upper-division science class about a particular sub-discipline or a real independent research project, other classifications or constraints will be more appropriate; these will be covered in the next section.

1. **Blue-Sky.** The first class of questions is the sort of question that humans have asked themselves over and over in the history of science, such as *"I wonder why the sky is blue."* We might call these "blue-sky" questions. Our chances of answering these big questions with a simple household science project are slim—but we might be able to find some small aspect of the problem in which we can make a small amount of progress. We also shouldn't discount the possibility that, in researching what others have found out about this big question, we might find smaller, more manageable questions that we can investigate. For example, in doing some background reading about why the sky is blue, we might find that the light scattered in the upper atmosphere ends up being polarized—that's something we can observe quite easily with a pair of polarized-lens sunglasses, and there may be a potential research project connected with that idea.

2. **Invention.** Another driving force in the history of science has been the effort of humans to make their life easier with inventions. Chemistry owes much to the metallurgists of the Bronze and Iron Ages, whose curiosity and inventiveness were

encouraged by the practical uses for these metals. Attempts to transmit information via electromagnetic waves led to the radio, to radar, and eventually to our understanding of the semiconductors that are an essential part of our computers and phones. The second class of "I wonder" statements I have my students come up with are ones where they consider what they might be able to do; an example is *"I wonder if I could* invent a more environmentally friendly soap?"

3. **Quantitation.** Once again, we can look at the history of science and see how important these questions were. Would Newton ever have come up with his laws of motion if Kepler hadn't had Tycho Brahe's excellent quantitative data on the orbit of Mars? Would Lavoisier ever have developed the idea of the conservation of mass if he hadn't decided that he needed to carefully measure the masses of all the reactants and all the products in his reactions? Scientists have always valued quantitative results over just qualitative descriptions, as it has been one way to ensure that findings aren't just subjective impressions. In my course, I require my students to design their household science projects so they can get quantitative results. A statement like *"I wonder if I can measure* the greenness of grass" may by itself not constitute a research plan, but there may be questions about grass, or other things, that we could answer with quantitative data about color.

4. **Comparison.** One sort of question that doesn't have a lot of resonance with the history of science is one that merely compares two approaches to a problem: *"Which is better* at wiping up spills, Brand A or Brand B?" These sorts of questions are easy to come up with, both because we frequently see these sorts of experiments in advertisements, and because answers to these questions could be very helpful when we go shopping. The biggest drawback to this sort of question is that it's hard to be very creative with it; remember, creativity is hard, and if it's too easy to come up with a question, that may very well be because other people have done similar things in the past. But if you are having a hard time coming up with questions, this kind of question might be a good place to start.

9.3 DIVERGENT THINKING IN A NARROWER, MORE ADVANCED CONTEXT

Rarely does a working scientist have the opportunity to consider all possible "I wonder" questions that he or she can come up with. If you're a plant physiologist, general questions like "Why is the sky blue?" are unlikely to help you come up with your next research grant proposal idea. Much of the time, researchers focus on incremental advances based on what other people in a given research field have been studying. If, however, a researcher wants to come up with something more creative, taking a fearless "I wonder" approach to more divergent thinking might still be in order. The difference is that we have to start in a way that keeps us within a certain discipline.

One way to keep our focus within a discipline is to first ask ourselves, "What are the popular big ideas and experiments in the field today?" Having identified four or five of these, we might ask the following kinds of questions:

1. **Questions about unexplained phenomena or discrepancies.** In the laboratories I teach, I often find science students assuming that incomplete agreement between the predictions of a hypothesis and actual experimental results is just "noise" or "experimental error." It is often worth wondering, instead, whether there is a good, and interesting, explanation for these discrepancies. In physics, discrepancies between prior predictions and experimental results, such as the Lamb shift, were often what led to new theoretical advances. Take a look at important experiments in your discipline and ask, *"I wonder if there is a way to explain this odd phenomenon...."* You could view these questions as essentially an advanced type of blue-sky question—you are wondering why something happens or why it is the way it is.

2. **Questions about applying big ideas to different contexts.** Ideas and hypotheses are usually most useful when they are applied to contexts that are fairly close to the context in which they originated. The idea of *orbits* was easy to take from Copernicus's ideas about the solar system and apply to satellites orbiting individual

planets within that system. But could it be taken to a *very* different context? Bohr tried to explain the structure of the atom by having electrons *orbiting* the nucleus. This simple model ultimately wasn't adequate, and we now think of electrons as being in *orbitals* rather than *orbits*, but Bohr's model was an important intermediate step in developing modern quantum mechanics. Divergent thinking questions along these lines might start out as *"I wonder if I could apply this idea to something rather different...."*

3. **Skeptical questions about popular hypotheses.** At any point in time, certain hypotheses become popular for explaining a lot of phenomena in a given discipline. A lot of times these hypotheses are appealing for how they can elegantly explain what is observed in nature, but they aren't always the only possible explanation. This doesn't mean that they are wrong—often these elegant hypotheses will turn out to be robust and long-lasting. But if we are trying to think divergently in order to be more creative, we might ask some "I wonder" questions along the lines of *"I wonder if there are other explanations for...."* In the history of science, one popular hypothesis was the idea of "spontaneous generation," the idea that living organisms could originate from non-living material or material from a different living organism. A dead carcass could soon become infested with maggots, and microorganisms could appear growing in pond water, and spontaneous generation was a popular hypothesis that could explain these phenomena. Creative scientists dared to wonder if there wasn't some other explanation, such as that the eggs of flies or the spores of microorganisms were the real origin of the organisms that seemed to appear "spontaneously." Creative experiments followed that provided evidence for the existence of eggs and spores, and now spontaneous generation is no longer a popular hypothesis.

9.4 CONVERGENT THINKING

After we have done our divergent thinking, we may have lots of interesting questions, but we can't pursue all of them, and there's a lot that has to happen between having a question and having an actual research plan. The point of divergent thinking is to have lots of ideas to choose from. The process of choosing a few ideas and developing them further is often called *convergent* thinking. In a group or an organization trying to solve a problem, converging is really the process of having different people unifying—converging—behind a single solution. If you're working by yourself, trying to find a problem, the word "convergent" may not be as apt. You will, however, need to combine your original "I wonder" question, your prior knowledge, suggestions and critiques from others, and information from the literature to come up with your research plan. Convergent thinking might be a pretty good description of this, too.

The important part of this second step is being critical. In a divergent thinking exercise, it is important to avoid being critical too soon, so that we don't prematurely discard ideas that might lead us in new directions. But criticism is vital to science, not only in examining the research others have done but also in deciding what research is worth doing. At this point in coming up with research ideas, we need to start asking *critical questions* about our ideas, and we can benefit from interacting with others.

In my class, I try to give students some feedback on the twenty "I wonder" statements in their first assignment, even if it's just a star by the best ones or a note pointing out which ones are the least original or most over-investigated. Students don't have to use this feedback; students may see potential in some questions that I miss. But, on the other hand, I have a lot more experience and knowledge in science and in household science projects! As we saw in previous chapters, science progresses because it is embedded in a social structure that preserves the discoveries of the past and builds on them. Professors are one way in which this heritage of knowledge and know-how is transmitted to students, so it makes sense to pay some attention to the advice they give.

Your peers can be another source of feedback; although they will probably have no more expertise than you, they may see problems and possibilities that you have missed. Now is the time for group discussion. Ask critical questions about your classmates' "I wonder" statements, and let them ask questions about yours. The kinds of questions we ask at this point are pretty easy, but they depend on which questions we're asking the questions about. For one of our blue-sky questions, the most obvious question about that question is "How do you think you're going to investigate that?" That's often a hard question to answer! It's pretty much the same for an invention question: "Do you have any ideas about how you might make something like that?" For one of the quantitation questions, one might ask, "What can you use that data for?" For questions in the other categories, you might well ask, "How do I get quantitative data to help me answer that question?" Finally, for almost any question, there is the one question that scientists are always asked: "Who cares?" Pretty soon, you are going to have to write a proposal for your research idea, and you'll have to come up with an argument about why your research is worth doing. Thinking about who cares, and why, are the first steps in assembling that argument.

Don't just stop with the general sort of critical questions I have given you examples of, though. Propose answers to the questions about your questions, and ask questions about those answers. As you go further, the questions and answers will be more and more specific to your research idea. Keep going. The deeper you go, the better your idea will be, and the better your project will be. Sometimes you will come to a dead end; you may realize that you just *won't* be able to explore a particular question with the resources you have available. And in that case, you will have to back up, think divergently again, and come up with something different. Maybe it will be a different one of your twenty original questions, or maybe your discussions will have given you a new idea. For scientists, skepticism is *constant*, and the best scientists never quit wondering if they are pursuing the right questions, or if there isn't a better way to answer the questions they are investigating.

9.5 VISUALIZATION

One strategy that is sometimes brought up in discussions of creativity is *visualization*. This word can mean a lot of different things to different people, but the general idea is that sometimes our brains work differently when they're thinking visually instead of thinking verbally. Your original twenty "I wonder" statements are words on a page—what would your questions look like if they were pictures, instead? It's hard to actually write a *question* in pictures, but it may be useful to try and picture answers. If you're trying to figure out why the sky is blue, it is helpful to draw some pictures of what we know about light. The sky only looks blue in the daytime; the light that allows us to see the blue sky must be coming from the Sun. Light often travels in straight lines—but if you try to draw a picture of light coming from the Sun at noon and the light of the blue sky coming from near the horizon, you will see that the light can't be just traveling in a straight line. In fact, that is at the heart of the answer for why the sky is blue; the light from the Sun is being scattered, and some wavelengths of light are scattered more than others.

Drawing pictures can also help as you get closer to coming up with a real experimental project. If you have a preliminary idea of how to do an experiment, it may be useful to sketch out the experiment. By doing so, you may end up seeing problems that need to be fixed, or you may see how the experiment idea can't possibly work. Either way, it is a good source of critical questions.

A more abstract way of using visualization to come up with critical questions is the concept map. The main idea in a concept map is that you can move upwards to more general theories or conclusions and downwards to other specific instances of these more general conclusions. Lateral movement on this map is also possible and is important in being creative—it is a form of divergent thinking.

How would this work? Perhaps your original question had to do with the greenness of grass. I've drawn a simple concept map that might be generated about the topic of "The Greenness of Grass" in figure 9.1. Grass is a specific example of the more general category of vascular plants, which also tend to be green. Why are they green? Because they contain chlorophyll, which is necessary for photosynthesis. In thinking this way, we have moved upwards to a more general concept, that of photosynthesis. Can we move back down to other more specific examples? Photosynthesizing bacteria and algae also can perform photosynthesis. By moving upwards to the more general topic of photosynthesis and downwards to other photosynthesizing organisms, we have actually moved laterally. Could algae

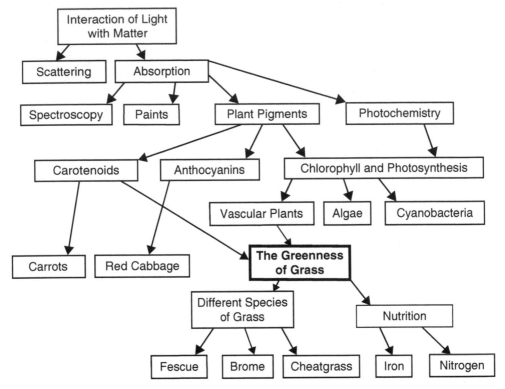

Figure 9.1. A concept map generated around the idea of "the greenness of grass."

or cyanobacteria be better experimental subjects for your research than grass?

But there isn't just one route through a concept map, either. Thinking more physically than biologically, the greenness of grass is connected with the general theories of light interacting with matter. If you can figure out how to measure the greenness of grass, you may be able to measure out the intensity of color of other objects. This may suggest other questions to you, and other avenues for research.

If you look at figure 9.1, you will see a lot of other potential connections. It is important to keep in mind that, as this is a tool for divergent thinking, you shouldn't really be too concerned with whether our concept map is *wrong*. I drew a line from carotenoids to the greenness of grass. Are there other pigments such as carotenoids or even anthocyanins in grass? I don't know. I'm not even looking it up to see if I'm "right" about that connection line. I'm just making a connection that I could investigate either in the literature or by experiment. There are other lines that I could have drawn but didn't. Can we use plant pigments in paint? There's a possible research project there.

A concept map can be used for convergent thinking as well, narrowing the scope of our research. The greenness of grass, for example, is really too broad for a simple research project—there are really too many variables. In the concept map, I have just highlighted two—the species of grass, and the amount of iron or nitrogen available to the plant. Either of these could be avenues for further exploration, either in the literature or by experiment.

9.6 SITUATING YOUR RESEARCH: THE SCIENTIFIC LITERATURE

After getting feedback on your original questions, asking critical questions, and finding possible answers to those questions, you probably still won't feel that you're ready to write a good proposal or to start doing research. You shouldn't. You need to connect your ideas with what other scientists already know about your questions and your proposed methods. Science is a social enterprise, and it has progressed because scientists don't just keep doing the same experiments over and over. The scientific literature has recorded what scientists have been doing for more than 300 years, and you need to find out where your research fits in to what other scientists already know and have already tried. The scientific literature may also give you ideas on how to better answer the question you have asked, or even suggest to you some new questions that you like better than your original question. Go back to chapters 3 through 7 to see how to get the most out of the scientific literature.

FOR FURTHER STUDY AND DISCUSSION

1. Write down twenty statements, five each in each of the categories mentioned in section 9.2. The Blue-Sky statements

should generally be phrased as *"I wonder why ...,"* the Invention statements should be phrased as *"I wonder if I could ...,"* the Quantitation statements should be phrased as *"I wonder if I could measure ...,"* and the Comparison statements should be phrased as *"I wonder which is better ... (or hotter or more efficient or ...)."* Try to make your questions as diverse as possible. It is certainly possible to develop several different statements from what starts out as a single idea, but this is defeating the purpose of the assignment. For example, you could ask *"I wonder if I can measure the greenness of grass," "I wonder if I could invent a device that measures the greenness of grass,"* and *"I wonder which grass is greener, mine or my neighbor's,"* but these are all centered on a single idea, the greenness of grass. This small-scale branching will be necessary in the next assignment, but for this assignment, the idea is to get twenty ideas that are as different from each other as possible. Now, twenty may seem like an awful lot, but you will find that a lot of them won't lead to workable projects. In my experience, twenty is the bare minimum needed to get students going on good, creative projects.

2. Choose four of your original twenty "I wonder" statements. Choose the ones that you think could actually turn into research projects. If you have come up with better questions since you wrote down your original twenty, you are welcome to use these as well. For each of the four "I wonder ..." statements, write down at least five critical questions and possible answers to those questions. The critical questions can either be ones that others have asked you, or ones that you have asked yourself. (Most likely, the in-class group work didn't last long enough for your group to have generated twenty critical questions for each person in the group, so you're going to have to come up with some questions on your own.) The later questions can be about answers to the earlier questions, or they can range broadly over different aspects of the problems. This assignment is just practice, though. By the time you generate your real research proposal, you will have to have asked yourself, and answered, many more questions about your proposed research.

3. Choose a concept from one of your original twenty "I wonder" statements. Generate a concept map that includes both more general and more specific concepts. Put enough branches in the map that, moving up and down, you move laterally to generate new "I wonder" statements. Generate five new "I wonder" statements that are not closely related to the original.

ADDITIONAL READING

Runco, M. A. *Creativity. Theories and Themes. Research, Development and Practice* 2nd edn (Elsevier, 2014).
Sawyer, R. K. *Explaining Creativity. The Science of Human Innovation* 2nd edn (Oxford University Press, 2012).

These are both standard texts on creativity studies. Despite some differences in organization and emphasis, both give a good background in the very complex, interdisciplinary field of creativity studies.

CHAPTER TEN

REFINING RESEARCH IDEAS AND WRITING A PROPOSAL

LEARNING OBJECTIVES

After reading this chapter, you will be able to:

- identify strategies for obtaining quantitative data
- identify what a Cartesian graph is, and explain why it is a good way to think about doing quantitative research
- anticipate obstacles or difficulties in a preliminary experimental plan, and propose ways to avoid or address these difficulties
- recognize that, while every granting agency may have a different prescribed format, it is important to follow the format completely
- describe the main kinds of information that all proposals have in common, despite differences in format between granting agencies

10.1 FROM IDEAS TO A PROPOSAL

In chapter 9, I wrote about how to use divergent and convergent thinking to get creative ideas for a research proposal, and I suggested that you try and situate your research by consulting the scientific literature. Ideally, you would probably want to do a lot of literature research and some preliminary laboratory research before starting to write a proposal. If, however, you find yourself in a one-semester class in which you have to come up with a research proposal and do the research, you don't have that sort of time. If you are in that sort of situation, you should choose research topics that you already know something about. This chapter focuses on how to develop research ideas and proposals in this context, although some of the same basic thinking strategies could be useful in other situations.

In the exercises for chapter 9, you developed some "I wonder" statements and generated some critical questions about those statements. The critical questions you asked should have helped you identify which ideas are most likely to give you a successful project. Now you need to ask even more critical questions and start coming up with concrete plans for doing experiments. The first half of this chapter will suggest key issues you need to address before you're ready to write up a proposal. You may have to explore these issues with more than just one of your potential research topics, though. In working out the practical details for one project, you may find it is unworkable, forcing you to abandon that idea and move on to something more workable.

10.2 PRACTICAL QUANTITATION

In my course, students are required to obtain quantitative data for their household science projects. If you are planning a research project that is done in a real scientific laboratory, the considerations you will have to address will be rather different, but some of the fundamental issues will be the same. Good quantitative data improves all research projects.

For a simple household science project, one of the biggest challenges students have is getting quantitative data for things they've previously evaluated in a more qualitative way. I'm going to work through some thinking for one common example, which involves quantitation of visual data. This isn't because you need to do a similar experiment for your project; I'm choosing this example because

I can use it to illustrate some of the mental steps and precautions you need to go through in getting good quantitative data. A good *general* strategy for thinking about quantitation is the following:

1. Consider what the physical *quantity* is that you usually just observe in a *qualitative* manner. For visual impressions, it is light; for the sourness of fruit, it would be acidity; for slipperiness it would be the amount of force required to accelerate an object a certain amount on a surface; for hardness, it would be the amount of force needed to make a dent of a given depth using a standardized tool.
2. Imagine what would be the ideal way to measure this, if budget and access to equipment weren't an issue. In some cases you may know the name of the scientific instrument you'd use and you may even have used it in a laboratory course; in other cases, you will have to imagine what such an instrument would be like, or you may find out from general reference works or the Web what sort of instruments people use to measure the quantities you're interested in.
3. Resign yourself to the fact that you might not be able to afford the fancy apparatus, and see if you can't get something that can measure some of the same parameters as the fancy apparatus. For example, there are expensive, well-calibrated hardness testers, but they function by applying known forces to an *indenter* of known geometry. Weighting down a marble with different masses and measuring the resulting indentation might serve to measure the hardness of candies or baked goods. One student did this to measure the hardness of fudge for a project in my course.

I'm now going to illustrate this general strategy for a problem that comes up frequently—getting quantitative visual data. Many students over the years have wanted to compare the effectiveness of different laundry detergents. You may have had the experience of putting a stained white t-shirt into the laundry, and you can still see traces of the stain on the shirt after the laundry is all done. Could a different detergent have done a better job? The obvious experiment is to try a bunch of different detergents and see which one *appears* to do a better job with the stain. But just saying the shirt appears more stained with one detergent is not quantitative data; we want to actually attach numbers to the degree to which the shirt remains stained.

What we have to remember is that a lot of our qualitative impressions are based on physical phenomena that can be quantified and that other scientists have figured out ways of measuring these phenomena. That spot where the t-shirt still looks a little brownish from the coffee stain is, in fact, reflecting less light in some regions of the visible light spectrum. In step 2 above, I suggested that you should imagine what you would use if budget and access to equipment were not an issue. If we had a reflectance spectrophotometer, it might be able to more precisely measure—assign a number to—the difference between the amount of light reflected by a faint spot and an even fainter spot, but in fact reflectance can be tricky to quantify if the light is being scattered in lots of different directions. We might be better off trying to shine light *through* the t-shirt, in a region of the spectrum where the t-shirt is more transparent—perhaps somewhere in the infrared, maybe—and see how much light is absorbed at wavelengths where the cloth fibers don't absorb. The stain we *see* is in the visible region of the spectrum, not the infrared, but the molecules that make up the stain may also have particular absorptions in the infrared.

Both of those suggestions, however, require a fancy spectrophotometer, which you may not have available. Is there some sort of technique that would be intermediate between simple qualitative description and careful quantitation using expensive laboratory instrumentation? Well, we in fact measure reflected light of different wavelengths all the time when we take pictures, and the pictures we take are now almost always digital—the light measurements have already been coded as numbers. We may have to do no more than get the numbers out of the pictures and make sure that, when we are comparing numbers from different pictures, we have taken the pictures in *exactly* the same conditions.

A lot of modern "point and shoot" cameras, including those on our phones, automatically adjust exposure based on the amount of light available, and this makes comparisons between photos mean-ingless if the camera selects different settings for the different pic-tures. If, for example, you took a close-up photo of a dark stain and a white t-shirt, where the stain filled up much of the area of the photo, the camera is going to adjust to make the stain lighter and the white t-shirt darker, and the actual numbers that are assigned to the light measurements will *underestimate* the difference in the amount of re-flected light. To use a camera to quantitate color or reflected light, you have to make sure your camera has a *manual* option, where the camera isn't automatically adjusting the exposure. You need to make sure that the distance between the camera and the subject is always the same, and that wrinkles in the t-shirt aren't creating shadows that are fooling your camera into thinking the stain is darker than it really is.

As for getting the numerical data out of the pictures, I won't give a lot of detail, because the software available for image processing is always changing. You'll have to do some background research! But a lot of good photo-processing software allows you to read off "RGB" (red-green-blue) values for particular pixels, or average val-ues for groups of pixels, and you can compare these numbers for different conditions and see if you can observe a trend. It is also often possible to convert the photos to black-and-white and get an overall value for the light reflected.

If all that sounds too complicated to you, there is a simpler way of quantitating something like the leftover stain on a t-shirt: com-parison. This is an old and respected way of characterizing color. Going back to the example of the coffee stain on the t-shirt: what you need is to create a *numbered standard scale* that shows various intensities of color, hold them up next to the stained t-shirt, and record what number on the standard scale is closest to the color of the t-shirt. The hardest part here is creating the standard scale, and it may require some experimentation to come up with a scale that is in the right range. In this particular case, the best would be to have pieces of t-shirt that were stained to different levels of faint brown. One might achieve these by staining the t-shirt samples with coffee of various levels of dilution. Because you'd be comparing the scale to a *washed* sample, you'd probably want to try it with fairly dilute coffee solutions. You'd need to keep track of the exact level of dilution (all made from the same pot of coffee) and expose the standards for the same amount of time before removing and drying. Once the standards were dry, you could compare them with the washed samples. Maybe Brand A detergent would give you the same color as a t-shirt soaked for one minute in a 1:100 dilution of coffee, while Brand B detergent would give you the same color as a t-shirt soaked for one minute in a 1:300 dilution of coffee. These quantitative results may not get your paper published in *Science* or *Nature*, but they are quantitative enough to give you something to graph, and they will allow readers of your study to get some notion of the size of any differences you might observe.

I led you through the example of quantifying visual observations in some detail because it may be one of the hardest things to quantify, even with expensive equipment, and because it seems to come up a lot in my students' research ideas. But the same general three-step strategy would apply to a lot of other challenging measurement issues:

1. Consider what physical quantity you usually observe in a qualitative manner;
2. Imagine the ideal way to measure this, if budget and access to equipment weren't an issue;
3. See if you can't come up with a way to at least crudely measure the key physical quantity.

10.3 USING QUANTITATIVE DATA

A further step in refining these ideas is to think about the interpretation of your quantitative data. The history of the scientific literature (see chapter 4) shows that the number of papers with quantitative measurements increased steadily from 1665 through to today. And, from the eighteenth century onward, one particular means of presenting that data has become common: the Cartesian graph. In a random sample of scientific articles from 1950 through the end of the twentieth century, 72 percent of the articles surveyed contained at least one Cartesian graph.[1] You are probably familiar with these graphs, even if you don't recognize the name. These are the graphs where you have some independent variable on the x-axis, and a dependent variable on the y-axis.

The increasing prevalence of these graphs in scientific papers reflects the increasing importance of mathematics to science. Mathematics often involves *functions*, and when you plot some measurable quantity against some independent quantity, you will very often get a function. For this reason, we often talk about graphs as if functions were exactly what we expected: "Plot the logarithm of the rate constant *as a function of* $1/T$." This means that $1/T$ will be on our x-axis, and the logarithm of the rate constant k will be on our y-axis.

One of the best examples of using plots of physical data to illustrate functions is simple physics, as shown in figure 10.1. The velocity of an object subject to the acceleration of gravity is $v = -at$. If time, t, is your independent variable on your x-axis, then the velocity, v, will be a linear function of t. The position, y, as a function of t will be given by $y = -at^2$, and y will be a quadratic function of t. This is Newtonian physics, and Newtonian physics has been a model for many scientists who have tried to make other disciplines more quantitative and rigorous. There has been a constant search for simple mathematical relationships between one variable and another, and very often the easiest way to see these relationships is with a Cartesian graph. Newton himself didn't employ Cartesian graphs, as they weren't yet invented, but the persistence and popularity of Cartesian graphs is evidence that they are well suited to discovering mathematical relationships in science.

The usefulness of Cartesian graphs is such that a lot of modern scientific instrumentation gives output in the form of Cartesian graphs. The simple UV-visible spectrophotometer, a common instrument in chemistry and biology labs, gives the absorbance of a sample as a function of wavelength or frequency. An oscilloscope, often used in physics and engineering to study oscillations in potential, gives a very high-speed view of potential as a function of time. I'm sure you can think of more examples that you have come across in your laboratory classes.

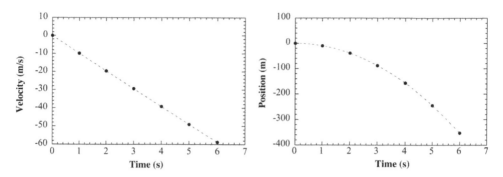

Figure 10.1. Plots of velocity as a function of time, and position as a function of time, for an object accelerated by gravity near the surface of Earth.

A household science experiment, however, may not be done with such fancy instrumentation. But you can still create Cartesian graphs, one measurement at a time, provided that you don't have to make the measurements too frequently. If you wanted to measure temperature as a function of time of an ice-water sample, you simply have to have a watch or clock that tells you what time it is, and then write down the temperatures you observe on a thermometer along with the times that you make the observations.

Is a Cartesian graph *necessary* for a science project such as the household science projects I have my students do? I will let your instructor decide that, but it is worth considering how a Cartesian graph might make your project better. Let's say you want to measure the effectiveness of different laundry detergents. You developed some way of quantitating how well a stain was removed, as we discussed in the previous section. How would you put this data on a graph? Your independent variable is the laundry detergent used—but there are no numerical values for the different brands of detergent, they're just names. If you have no clear numerical values for your *x*-axis, you could construct a bar or column graph, which just shows cleaning effectiveness for the different detergents.

Could you do better? If you had some *meaningful* numbers associated with the different detergents, then you could plot the detergents in a Cartesian graph. To do this, you would have to have some hypothesis about what makes some detergents better than others. Is there a difference in the pH of aqueous solutions of the different detergents? If so, you could make a graph of cleaning effectiveness as a function of pH. You might find from plotting the graph that pH doesn't matter, or that all the pH values for the different detergents are so close to each other that pH isn't a relevant parameter. But at least now you have made a hypothesis and tested it.

Of course, the difference in the different laundry detergents might be due to different amounts of all sorts of different chemical components. One big obstacle in testing the effect of different components is that most detergents have proprietary formulations, which means that the manufacturers won't tell you exactly how much of what is in the detergent. Depending on the materials available to you, however, you might be able to learn more. If you could find a source of some common ingredients of laundry detergents, you could formulate a series of detergents in which the quantity of only one component is varied. You could also try adding a single additive to a laundry detergent that doesn't already do a good job. People used to do this a lot in years past, adding washing soda, special laundry bar soaps, bluing, or bleach to their laundry. Many of these additives are still available commercially, and you could try adding varying amounts to a simple laundry detergent. You'd then have an independent variable, the amount added, and you'd have an experiment that tells you something a little more fundamental about washing laundry than just comparing brands.

What I hope you've seen from this example is that some restrictions on your project—requiring quantitative data and a Cartesian graph—might actually end up making your project better, because they force you to think about the underlying science more precisely. This is an important part of the "refining research ideas" that is mentioned in the title of this chapter, and it represents a continuation of the convergent thinking process we started in chapter 9. Yes, creativity comes from setting your mind free to explore lots of different ideas, but it also involves subjecting those ideas to criticism, asking how they could be better. In some cases, you might have to abandon an idea and go back to divergently thinking of other ones—but that is all part of the creative thinking process.

10.4 WHAT ABOUT STATISTICS?

The ideas of probability are hundreds of years old, and they have been applied to all sorts of natural and social sciences. Many of the theoretical underpinnings of statistics were developed by late-eighteenth- and nineteenth-century mathematicians. The widespread use of statistics in interpreting scientific data, however, is pretty much a twentieth-century phenomenon. A lot of early physical science was not done with any statistical analysis. This was possible because, in simple physical systems such as the orbit of planets, the number of variables affecting the observed position of the planets is relatively small. In the emptiness of space, there aren't winds blowing the planets around, and, aside from some distortions due to fluctuations of the atmosphere and small effects of gravity, the light we use to determine the positions of the planets generally travels in straight lines. So, when Isaac Newton came up with his equations of motion based on the motions of the planets, he could come up with models that matched the data pretty well, without having to worry too much about random variations and error.

When the physical sciences moved from the planets down to the relatively messy surface of the Earth, experiments could be designed to minimize the number of variables that might affect the result. By controlling the conditions of experiments carefully—so-called "one-factor-at-a-time" or OFAT experiments—nice Cartesian graphs could be constructed that showed clear mathematical relationships, such as those shown in figure 10.1. Real data might be a little noisier than the figures shown; there might be some deviations from individual data points to the fit mathematical function. The conclusions, however, are pretty evident as long as the deviations are small. The strategy of just testing one factor at a time still works pretty well for a lot of studies in the physical sciences, and many simpler systems in biology as well.

In the social sciences or the study of whole organisms or ecosystems, however, the systems being studied can't really be taken apart to isolate individual variables. If you want to study the relationship between high school graduation rates and average annual incomes for different states in the United States, there will be a lot of

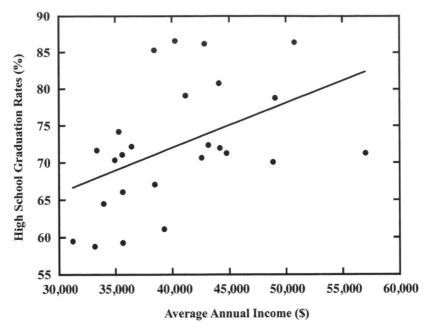

Figure 10.2. A plot of high school graduation rates as a function of average annual income for twenty-five states, with a least-squares linear fit suggesting a possible correlation.

uncontrolled variables such as movement of people between states and the presence of prosperous industries affecting the data. You won't get a nice fit like the one in figure 10.1; it will look more like figure 10.2. If you want to figure out how strong the relationship is between high school graduation rates and average annual income, and you want to make sure the data aren't just a random assortment of points that just *look* as if there is a relationship, you will have to use statistics.

I can't teach you statistics here, because that usually requires a whole semester and an instructor who understands the subtlety of different statistical measures better than I do. The reason that I include it here is that it is something you *may* need to worry about in proposing a research project. There exists a whole field of statistics called *design of experiments*, pioneered by British statistician Ronald

Fisher, where the use of statistics is anticipated and incorporated into how experiments are designed. If you are venturing into areas where there will be a lot of variables, such as clinical trials for new drugs or treatments for disease, it will have to be something you look into. For the simple at-home science projects that I have my students do for the purposes of learning how to propose and write about scientific experiments, however, I do not require careful statistical design or analysis.

Another thing to keep in mind is that you *should* eliminate uncontrolled variables whenever you can. It is true that while a wise use of statistics can objectively determine when a relationship exists between a controlled variable and the observable, people, including scientists, will always be readier to accept your hypothesis when your data don't look too "noisy." People are far more likely to believe the mathematical relationships shown in figure 10.1 than the one suggested by figure 10.2. Spending time anticipating experimental errors and uncontrolled variables *before* doing your experiment can save you from having to do a lot of statistical analysis afterwards, only to find that your results are not statistically significant.

10.5 ANTICIPATING PROBLEMS

After you have thought through the how and why of your quantitative experiment, it's a good idea to imagine the experiment in more detail. There are always lots of obstacles that you won't have anticipated, but it is good to get these down to a minimum *before* you write your proposal. If you think about all the potential problems and address them in your proposal, more experienced scientists will be less likely to reject your proposal because it has obvious obstacles.

One good way to imagine your experiment is by sketching it. It doesn't have to be a work of art; it just needs to be a useful thinking tool. Figure 10.3 shows, for example, a sketch illustrating how one might measure the maximum height of a muffin. What are the problems here? One of the problems is that the ruler is not right next to where the height of the muffin is at a maximum. Depending on the

Figure 10.3. A rough sketch showing possible difficulties in accurately measuring the height of a muffin.

angle at which you look at the muffin and the ruler, you might get different measurements. (The official scientific name for this problem is *parallax error*.) How do you solve this problem? There are several ways to do this, but it's something that would be good to think about before describing your experiment in your proposal. If you don't include some way of solving this problem in your proposal, experienced scientists might reject your proposal because there's an obvious source of error that you haven't guarded against.

As we discussed in the last section, it's also a good idea to anticipate uncontrolled variables and noise in your data. The best way to do this is to think divergently about all the different variables that might affect your observable. You may be concentrating on just one variable, but others may be lurking, potentially making your results meaningless. For example, if you were measuring the greenness of grass in two different plots, and trying to see the effect of nitrogen fertilizer, you could get an entirely erroneous result if the soil in one plot contained more iron than the soil in another plot. With multiple plots for the conditions of no fertilizer and fertilizer, you could use *statistics* to figure out if the grass is greener *on average* when fertilizer is applied, but then you have to think carefully about statistics. As I suggested in the last section, controlling variables is sometimes easier than hoping that the statistics will allow you to make a compelling argument. If you were to dig up and mix thoroughly the soils from the two locations, and put this mixture back into the two plots, you'd be more certain that nutrients other than nitrogen were the same in the two plots. Of course, even with this precaution, there would be more variables to worry about—could the process of digging up and mixing the soils affect organisms in the soil such as bacteria, fungi, and nematodes? What if there are pebbles in the soil that can't be distributed evenly throughout the soil mixture? What if the roots of the grass go deeper than you dug? These are the sorts of questions that make good science difficult to do, but it is much easier to ask them in advance and try to learn from them than to wait until you spend time collecting meaningless data.

10.6 WRITING THE PROPOSAL

When people think of research proposals, they often think of proposals submitted to government agencies to get money for academic research. Those are the sorts of things your professors worry about. But proposal writing is much more widespread than that. Scientists who work in industry have to propose new research directions to their superiors before they can launch into these new research ventures. New companies are founded based on proposals submitted to venture capital firms who finance the new companies. Even outside science, people write proposals to get money for social service programs and educational grants. In today's world, many people have to make the case that money is worth investing in projects that won't have an immediate cash return on investment. So, even if you're *not* planning on becoming an academic scientist, this may be the most applicable writing you'll learn about from this book!

There is no standard format for writing a proposal. Or, rather, there are hundreds of standard formats for writing proposals. Every organization that gives out money for research is going to have its own format, and very often such organizations require very strict adherence to that format. Some government agencies go so far as to specify which fonts are acceptable and how big the margins have to be; this is mostly to keep people from trying to pack too much detail and background information into their proposal by making the fonts and margins very small.

There are, however, some common elements to research proposals. One could classify these into a few categories:

1. Who are you?
2. What do you propose doing?
3. Why should we believe that you could do this?
4. Why should we give you the money to do this?

The questions asked by granting agencies will often not be so direct and not organized particularly well into those categories. Biographical information will be used to answer questions 1 and 3. The technical part of your proposal will be used to answer 2, 3, and 4. Information about the institution you're part of will be used to help answer 1 and 3. But the intent of all of the specific questions and requests for information is to give the people with the money confidence that you can be trusted to use the money in a productive way.

In my course, I don't actually give students money. That's because I have none to give. But I do need to be convinced that they have a good, safe, feasible project before I allow them to start work on it. And I require the following information, *all clearly labeled*:

1. Project Title
2. Name and Title of Principal Investigator
3. Institution
4. Project Summary
5. Background
6. Experimental Plan

7. Institutional Support
8. Budget
9. Institutional Approval

The *Project Title* has to give the reader a clear idea of what you are proposing to investigate. I find that students tend to make their titles too general. Keep in mind that the organization to which you are submitting the proposal may have lots of similar proposals, and you want to catch the attention of the reader with some specifics. On the other hand, don't make the title so long that people don't want to read it.

The *Principal Investigator* is you, and so it's your name and title that are needed here.

The *Project Summary* is like an abstract, but its intended audience isn't just scientists. It includes executives and other people who wear suits and hand out money, so don't make it too technical and be sure to emphasize the benefits of what you will discover. But don't completely leave out your experimental approach—"the suits" are sometimes people who used to be scientists, and they're going to want some sense of how you intend to accomplish your research goals.

The *Background* goes into more depth on the importance of the problem, as well as what is already known about it. It is similar to the *Introduction* of a scientific article, in that it will start with broad statements about the importance of the questions you are addressing with your research, cover what is already known about the topic, and finish with a statement about how your research will help fill in existing gaps in knowledge. The middle section will include what you learned from your background research (chapter 4) as well as any personal knowledge you may have of the subject.

The *Experimental Plan* is where you need to explain in detail just what you intend to do. All of the thinking you did to anticipate possible problems is relevant here, and you need to address what you are doing to avoid these problems. People reading proposals are often more experienced than the people submitting them, so they can more easily imagine potential problems, and they may criticize your proposal for failing to anticipate and address these. This is where you really need to convince the reader that you can success-fully do the research. Don't leave out details because you assume that the reader will assume that you'll do everything right. I require rewrites on experimental plans where students just write that they will measure the maximum height of muffins, without explaining how they will try to eliminate parallax error. The *Experimental Plan* is also where you need to reassure the reader that you've thought about the safety hazards of the experiment and will take proper precautions.

The *Institutional Support* is another opportunity to convince the reader that you can accomplish what you have proposed. This is something required of academic researchers who submit proposals to government agencies for funding; the government agency needs to know that the researcher has the laboratory space necessary as well as any instrumentation that's needed for the research but that is not funded by the grant. In the case of an at-home student proj-ect, this would include equipment like dishes, ovens, washing ma-chines, blenders, sinks, thermometers, and kitchen scales that might be used in the research. These things all fall under the category of *capital equipment*. Capital equipment, for our purposes, is anything that's not used up in the course of the research. If you already have flour or paper towels available at your home, don't list these as capital equipment, because they're going to get used up. They are *expendables*, and should be saved for the next section, the budget.

The *Budget* is a table of all the expendable materials used in the experiments, as well as any additional capital equipment that you will have to purchase. As I noted above, I don't actually have money to give my students, but I still require a budget. I have a limit for how much students can spend on their projects, and the budget allows me to make sure they're not exceeding this limit. Also, putting together a budget is an essential part of writing a proposal. In university research projects with government funding, the largest part of the budget is usually personnel costs; principal investigators and their students are paid for doing the actual work on the experiments.

The *Budget* should be a table with individual items, the cost per unit for the item, the quantity needed, and the extended cost. See table 10.1 for an example. Budgets don't have to be accurate to the penny. I constructed table 10.1 with information I could get from the Web, including the conversion of teaspoons (tsp) to ounces (oz) for cardamom! You may end up paying a little less or a little more when you actually go to buy the stuff, but you can at least get an approximate budget put together without having to go to the store. (I just invented the recipe, though—don't bake this at home!)

Check to make sure that everything you mention using in your *Experimental Plan* is accounted for either in the *Institutional Support* or the *Budget*. Otherwise the reader will question whether you can actually accomplish your research objectives within budget with the capital equipment you have available to you! You should also make sure that nothing appears in both places.

The *Institutional Approval* is also a very typical part of proposals in the real world. If I propose to do research in my laboratory, I need quite a few people at my university to sign off on my proposal. The Health and Safety Office needs to sign off that they're okay with what I am doing; the department chair has to sign off that he or she is okay with my promising to do this research, especially as it may take away from my availability to teach courses; the dean of my college needs to sign off, for similar reasons; the financial folks need to make sure I haven't misrepresented the university's policies on

Table 10.1. A sample budget for a simple baking project.

Item	Cost/Unit	Quantity	Extended Cost
Flour	0.40 /lb	0.50 lbs	0.20
Sugar	0.55 /lb	0.25 lbs	0.14
Eggs	0.25 /ea	2	0.50
Milk	1.00 /qt	0.25 qt	0.25
Cardamom	7.11 /oz	0.018 oz*	0.13
Thermometer	5.99 /ea	1	5.99
Subtotal			7.21
Tax (thermometer only)			0.51
Total			7.72

*0.018 oz = 1/4 tsp

indirect costs before they sign off; and then there are even more signatures. For a household science project, you will also need people at your "institution" to sign off, because your use of the kitchen, the backyard, the garage, the oven, the sink, or whatever, might inconvenience or even endanger them. I have my students copy the following statement at the end of their proposals and have all affected parties sign below the statement:

I have been fully informed of the nature of the research to be conducted by (Principal Investigator name) at this location, and I grant full approval and support to this research.

FOR FURTHER STUDY AND DISCUSSION

1. Sketch an outline for an experiment. This sketch can be a flow chart of what will be done or a particular apparatus or method to be developed. Trade your sketch with a classmate. Look for possible problems or obstacles with one another's sketches, and propose solutions to these problems.

ADDITIONAL READING

Hofmann, A. H. *Scientific Writing and Communication. Papers, Proposals, and Presentations* (Oxford University Press, 2016).
    A large part of Hofmann's book is devoted to a detailed consideration of all aspects of proposal writing.

CHAPTER ELEVEN

THE LABORATORY NOTEBOOK

LEARNING OBJECTIVES

After reading this chapter, you will be able to:

- explain the reasons for keeping a good laboratory notebook
- identify the rules which always must always be followed when keeping a laboratory notebook, and the rules which may be different for academic, industrial or government contexts
- explain why some rules are always followed and why some rules depend on the research context
- identify advantages and disadvantages of electronic laboratory notebooks

11.1 THE EVOLUTION AND IMPORTANCE OF THE LABORATORY NOTEBOOK

You have probably kept a laboratory notebook for laboratory science courses in high school and college. You may think you know all about how to do it. Like other forms of writing, however, there's always room for improvement. Instructors in science classes are sometimes so busy making sure their students can successfully run experiments, collect data, and interpret that data correctly, that they have little time to focus on *all* of the aspects of a good laboratory notebook. If your instructors *have* had the time to carefully instruct you on how to keep a good laboratory notebook, and have provided you a lot of useful feedback on your own notebooks, then you are fortunate!

The laboratory notebook has been an important part of scientific research for a very long time. Leonardo da Vinci's notes are the primary way in which we know the breadth of this 15th-century great's accomplishments in science and invention, even though most of these were not originally in a bound notebook. These notes are hard to read; the notes were not only written in a script that was written right-to-left (perhaps due to da Vinci's left-handedness) but also with unusual punctuation and spelling. Leonardo himself is said to have lamented over the disorganization and confusion of his notes, as he wished to have many of his discoveries and inventions published. If his notes had been better organized and more legible, such publication might have happened during his lifetime, instead of long after his death!

Leonardo's notes were well known in the 16th and 17th centuries and may have been a model for future scientists. We also know that many of the members of the Royal Society, whose history we reviewed in chapter 4, kept notebooks when they made their observations and did their experiments. Notebooks of Antoine-Laurent de Lavoisier, the French scientist often thought of as the founder of modern chemistry, have provided excellent evidence of the evolution of his thoughts and methods.[1]

Like the scientific journal and the scientific article, the laboratory notebook has evolved to better serve the needs of those who use it. As a student in a laboratory course, your primary interest in putting together a laboratory notebook may have just been to get a good grade from your laboratory instructor. But outside of a classroom setting, there are many other reasons for keeping good laboratory notebooks, and understanding these reasons will better help us understand why laboratory notebooks have to be kept with care.

REASONS FOR THE LABORATORY NOTEBOOK:

1. *To aid the memory of the researcher.* First and foremost, you want to record everything—ideas, procedures, data, observations and interpretation—with as much detail as you can. Every scientist has thought, at many points in their career, that they would remember the important facts of an experiment, only to find that there were gaps in their memories only a few days later. Often, it isn't a few days later that the facts need to be remembered; lines of research can be put aside for months or years while something more promising is investigated. When it's time to go back to the original line of research, much wasted time and re-experimentation can be avoided by having detailed notes of procedures and results.

2. *To communicate procedures and data to other people within a research group.* Scientists rarely work alone in laboratories these days. Other people in the same research group will want to build upon your successful experiments, and they'll benefit from very complete descriptions of methods and materials. Senior scientists will want to see your procedures, data and observations because their more experienced minds may see particular details as significant, even if they don't mean much to you.

3. *To provide a basis for published or summary reports.* Although it is most often students and professors in academia who aim to publish papers in the scientific literature, industrial scientists often have to present their data to people higher up in the company in the form of reports or memos that summarize findings. More complete and better-organized lab notebooks make the writing of papers and reports easier.

4. *To provide a defense against accusations of fraud.* Unfortunately, not all the papers that make it to the scientific literature measure up to the standards of honesty described in chapter 2; some describe work that wasn't performed as reported, or that didn't give the results that were reported. These cases of fraud are usually uncovered when others are unable to reproduce the results of a given study. Usually there isn't an immediate accusation

of fraud after publication; years can go by before somebody tries to replicate the experiments described. The original laboratory notebooks in which the data were taken are the best defense against accusations of fraud. If there is no record of the experiments reported, or if there is data that is different from the data that is reported in the final paper, questions about the scientist's honesty will be raised, and a young scientist's career could be ruined.

5. *To document priority of invention for patents.* In industry, millions of dollars can be at stake when a company files for a patent. In some cases, other companies will be filing simultaneously for similar patents or may challenge some patents because they weren't the first implementation of a particular idea. In some of these cases, it is important to determine who had the idea for an invention first, and who actually made a working prototype of an invention. Very often it is a laboratory notebook that provides evidence for this priority of invention. For this reason, industrial notebooks are often signed and dated by the lab worker as well as by a witness who understands the notebook but who would not directly benefit from a patent that might result from the work.

6. *To provide historians of science with a more direct record of the evolution of ideas.* Scientists don't usually consciously think about this when writing a notebook, but laboratory notebooks can provide rich detail of scientists' thinking and methods that goes beyond what ends up in brief scientific papers. Some scientists may be horrified to think of historians perusing their messy and incomplete notebooks—it may be wise to assume that your notebooks will someday be perused by historians!

The most basic physical requirements for the laboratory notebook are dictated by reasons 4 and 5. If the notebook is to defend against accusations of fraud, or show the priority of an invention, it has to show a clear, legally acceptable record of what was done on a certain date in the past. Anything that detracts from the permanence of this record makes the notebook less useful for these purposes. Therefore:

1. All entries must be written in ink, preferably black ballpoint pen. Pencil can be erased; other pens may fade or bleed if the notebook gets wet.
2. All entries must be in a bound notebook, not a loose-leaf or spiral notebook. It should be difficult, if not impossible, to remove pages without leaving some trace. For this reason, notebooks with pre-numbered pages are preferred.
3. Notebooks should be of sufficient quality to be stored for a number of years, in case of patent challenges or fraud investigations.

The idea of the laboratory notebook as a potential legal document also affects how we record data in it:

1. At the very least, every entry into the notebook must be dated. Avoid using number-only dates, as these can be misinterpreted. What is 3/10/2018 in the United States may be 10-03-2018 in Europe. It is better to write "March 10, 2018."
2. For notebooks that are used in industry, every entry should be signed by the person doing the work, and, in most cases, signed by a witness who understands the work described in the notebook entry.
3. All data and procedures are entered *directly* into the notebook in ink, not copied from "scratch paper" notes done on loose sheets. It may be helpful to think of the notebook as being a *witness* to the experiment, testifying that it was there and it got all of the information directly and immediately.
4. Any mistakes should be crossed out with a single line, so that what was crossed out can still be read. All cross-outs should be initialed and dated. Even better is an explanation of the cross-out, so that people looking at the entry in the future will understand that you're not trying to rewrite history.

Academic researchers—students and their supervisors—would probably benefit from following the same rules as industrial researchers. In practice, this often doesn't happen. But academic research can often result in patents, too. And even an accusation of scientific fraud or misconduct can damage a scientist's reputation. Better safe than sorry!

The basic mechanics of making entries, however, is the easy part; much harder is the habit or discipline to make complete, clear, well-organized entries. I know that some instructors in particular laboratory classes may have given you very exact instructions of what to write where, but in the wider world, there aren't such strict rules. But once again, if we think of the *reasons* for the laboratory notebook, we might have a better idea about how to write it.

When motivating yourself to write clear and complete entries, think about the first three reasons I gave for the existence of laboratory notebooks: reminding yourself what you did and what you observed, providing information to other people in your lab group, and making it easier to write papers and reports. Many times, you might be able to convince yourself that a particular piece of research won't result in a patent, or it won't be important enough for anyone to investigate for fraud. But your memory will *always* fail you to some extent, and others will be frustrated if your failures in memory mean a delay in a publication or the reinvention of a method. Not only that, but you never know when an accidental discovery *could* result in a major discovery or patent! Therefore, good laboratory notebook keeping should be a habit—a virtue—that you cultivate at all times.

11.2 THE FORMAT OF A NOTEBOOK ENTRY

In laboratory classes, your notebook frequently doesn't really need to serve as an aid to your memory for more than a semester. But in more extended research projects in medicine, academia and industry, data taken years ago can be relevant. It is not just enough to write down the data, but also to write down the purpose of doing the experiment or observations, the complete methods, and thoughts about what the results mean. I have had too much experience looking at my own notebooks from years ago and wondering, "Why did I do that?" or "Why did I follow up this experiment with that one?" That's because I described the experiment and the results, but didn't write much (if anything!) about my thinking in deciding to do certain experiments or what I hoped to gain by doing them.

One suggested format for a laboratory notebook entry[2] is similar to the format of a scientific paper, with sections entitled *Introduction* or *Purpose, Experimental Plan, Observations and Data, Discussion of Results,* and *Conclusions.* These roughly correspond to the *Introduction, Methods, Results* and *Discussion* sections of a standard scientific paper, but they're not exactly the same, as we'll discuss below.

The *Introduction* section may be better called the *Purpose* section, in that it's primarily there to remind you, or someone else who later reads your notebook, what the reason was for doing the experiment. This sort of introduction is not likely to have as many references as the *Introduction* section of a scientific paper, nor will it serve the same rhetorical purpose.

The *Experimental Plan* section is in some ways analogous to the *Methods* section of a scientific paper, but it outlines what you *plan* to be doing, rather than what you actually end up doing. The actual masses of reagents used and changes to the procedures that happen once the experiment is started will be recorded in a different section, or perhaps in a separate column of the same section. The purpose of the *Experimental Plan* is to force you to think through potential issues carefully, and to provide you with a list of you what to do next when the actual experiment is in progress. Drawings of planned setups of experimental apparatus are very useful, and can help you identify the equipment you need to round up before beginning the experiment, as well as to see potential problems. The amount of detail you put into this section is something you will have to decide. Too much detail may mean too much time spent writing long lists of individual actions that any person experienced in laboratory science would assume anyway. Too little information may risk leaving out important details that *can't* be assumed. This is one of the aspects of writing a laboratory notebook that you will get better with experience.

Although the *Observations and Data* section does, to some extent, parallel the *Results* section of a scientific paper, it includes a lot more. It will include some information that may ultimately end up in the *Methods* section, as you record the exact amount of chemical reagents used and their source; apparatus used, its condition, and its means of calibration; any organisms used, and their source and condition. But it will also include the data that ends up in the *Results* section of a published paper. This data may very well be much more detailed than what ends up in the paper; the paper may have averaged data presented, whereas the notebook will have all individual measurements. It is best to take down data in a neat, organized table if you are writing down observations or measurements by hand.

It is now frequently the case that data exists as computer outputs, computer files, or digital photographs. Here you have to make a decision: do you paste in the printed output, do you collect the output in a separate notebook or binder and refer to it in your notebook, or do you just store it digitally and refer to it in your notebook? The volume of data may help make your decision for you; my own students frequently have hundreds of digital micrographic images, and it makes no sense to print these all out, especially when we often only look back to selected ones. When my students regularly take spectra of chemical products, it makes sense to print them out and keep them in separate binders, well labeled with the date, sample, and other information, as trying to paste them all into the notebook would break the binding of the notebook. But an occasional spectrum, pH vs. time plot from a chart recorder, or some other sort of data not taken often enough to warrant its own notebook is best pasted right into the notebook where it can easily be found.

Both the *Discussion of Results* and the *Conclusions* are for you to think about what the data mean and what you can conclude from it. For some simple experiments, perhaps only *Conclusions* will be necessary, but do not underestimate the value of engaging in a written *Discussion* of the results. Sometimes, in trying to express your thoughts about data in clear sentences, you will realize that your thoughts don't make as much sense as you thought they did while you were preoccupied with the experiment. Making graphs of numerical data may also help you see trends that you hadn't noticed previously, or may suggest future experiments that need to be done to establish a functional relationship.

11.3 THE LABORATORY NOTEBOOK IN REAL LIFE

Thinking too much about how a laboratory notebook *ought* to look could potentially prevent us from allowing our laboratory notebook to fulfill its key functions. Very strict requirements for format sometimes lead students to think that the only way of doing it right is to do a rough draft on loose paper, to be recopied more carefully into the laboratory notebook. But this is *worse* than doing a sloppy job, as one of the primary reasons for the lab notebook is to provide an immediate, direct recording of what goes on in lab. In other cases, the demands for detailed lab notebooks lead students to assume that they'll never be able to meet expectations, and they'll give up on ever doing a better job. That's not the right lesson to draw from this chapter, either. By all means, the most important thing is to be writing down as much as you can as you are doing the experiment!

I personally think that a wise philosophy for the lab notebook is this: *Do the best that you can; it may never be perfect; keep trying to improve.* One thing I find helpful is to include notebook-keeping in the time you budget for laboratory. Don't just estimate how long it will take to do an experiment, but how long it will take to introduce and plan an experiment in the notebook, do the experiment while carefully taking notes and data, and write a discussion and/or conclusion about the experiment after the experiment is done.

11.4 ELECTRONIC LABORATORY NOTEBOOKS (ELNS)

The idea of an Electronic Laboratory Notebook has been around for many decades, but they have only come to be used widely since 2000. One of the obstacles to earlier adoption of ELNs was the concern of many that an "electronic-only" notebook wouldn't be accepted by regulatory agencies such as the Food and Drug Administration (FDA), or by the courts in a patent dispute. However, regulatory changes in both Europe and the United States now mean that, if proper "electronic signatures" are implemented within the ELN software and database, electronic records do have the same legal standing as paper records. In the early 2000s, companies employing ELNs often kept "hybrid records," with data entry being made by ELN, but printed copies of ELN pages being printed out, signed and dated by hand. But that practice is fading as the use of electronic signatures becomes more and more commonplace in all areas of society.

Some advantages of ELNs are the following:

1. If data are stored in a central database, many scientists in an organization (a company or academic research group) can have access to experiments and data that other scientists have performed and collected.
2. Electronic keyword searches can be done, rather than relying on an often-unreliable Table of Contents written by the writer of the laboratory notebook.
3. Repetitious experimental plans for similar experiments can easily be copied and pasted so that entries for every experiment have all the details necessary to duplicate the experiment.
4. The ELN software can help to guide the scientist, prompting the scientist for all required information and placing it in the proper format.
5. Electronic outputs from instruments often can be put directly into the ELN in digital form, without tape or glue, avoiding the problem of breaking the binding of the notebook when too many printouts are pasted in.
6. With proper storage, digital records can outlive paper records; storing a copy off-site means that the data is preserved even if a fire completely destroys a lab and all the paper records.

But there are some disadvantages, too:

1. Because of the need for secure electronic signatures and special graphics capabilities, you can't just take any word-processing or spreadsheet application and make a legally-acceptable ELN with it.
2. There are lots of different vendors of ELN systems (around 30, at last count), and all of the products are a little different, and each has its strengths and weaknesses.

3. There's no guarantee that a vendor will continue to support your product or even stay in business, and not all ELN systems work on all platforms.
4. A robust system will often require building a database on a centralized server to maintain the records, and this often means more cost and personnel.
5. Individual scientists in an organization may have note-keeping styles that don't work well with the centralized ELN system that the organization has chosen.
6. They cost much more money than a paper notebook and a pen.

Because ELNs do cost more than paper and pen, they are most popular where the advantages are worth the cost. Some of the first adopters of ELNs were pharmaceutical companies, who already were working under very strict requirements from the FDA for keeping records, and already had much of their instrumentation networked to facilitate the collection and storage of data. As of May, 2013, the Royal Society of Chemistry's *ChemistryWorld* was reporting that a number of academic institutions were starting to adopt ELNs, but most of these are major research universities or institutes. Ultimately, the decreasing cost of technology may result in ELNs being ubiquitous, even in high-school chemistry labs. But for now, there's still a future for the paper notebook.

FOR FURTHER STUDY AND DISCUSSION

1. Establish a local standard for laboratory notebooks with your classmates. Is this standard designed for use in industry, academia, or some other context? What disagreements do you have about the notebook standards? Do these have to do with the different sort of research projects that different students are imagining? How can you make the standard flexible enough to accommodate different kinds of research while still maintaining the most important features of laboratory notebooks?

ADDITIONAL READING

Kanare, H. M. *Writing the laboratory notebook*. (American Chemical Society, 1985).
    Although this is primarily concerned with chemistry laboratory notebooks, there is good advice here for all scientists.

CHAPTER TWELVE

SCIENTIFIC WRITING: GRAMMAR AND STYLE

LEARNING OBJECTIVES

After reading this chapter, you will be able to:

- identify the tense in which a sentence is written
- identify whether sentences are written in the passive, active, or imperative voice
- rewrite a sentence in the active voice into the passive voice, and do the reverse
- identify strategies for improving problem sentences

12.1 TENSE AND VOICE

Before we get into the actual process of writing, we need to consider general questions of grammar and style. I'm not going to give you a complete course in English grammar; I expect you already have some grasp of it. But a lot of the grammatical issues—they might even be called controversies—in science writing revolve around how we use verbs, so it is probably a good idea to review verb *tense* and *voice*.

Tense indicates whether a sentence describes something that happened in the past, is happening now, or will happen in the future.

You would think that we should only need three tenses for this, but in fact we have more, some with subtly different implications. For example, the *perfect* tenses can add an additional sense of time. I like to think of the *past perfect* as the "past of the past"; it's used in sentences like "By the time of the Roman Empire, the Babylonian kings had been dead for more than 500 years." The Roman Empire is the past to us; to people in the Roman Empire, the Babylonian Empire was the past. In writing about both of these past times, one further in the past than the other, the past perfect comes in handy. Fortunately, you rarely have to use the past perfect (or other perfect tenses) in scientific writing; it's best just to stick to the past, the present, and the future.

Voice can mean a lot of different things when we talk about writing, but we are using it here mostly to distinguish between the *active voice* and the *passive voice*. The active voice is usually the more direct way of writing a sentence: *I filtered the sample.* "I" is the subject, the person or thing doing the action, which here is indicated by "filtered," the past tense of the verb "to filter." "The sample" is the direct object, the thing being filtered.

In the passive voice, we put "the sample" up front as the subject, and modify the verb by adding some form of the verb "to be" (i.e., *is, are, were, will be*, etc.): *The sample was filtered.* Notice that there is no object. We can add a *prepositional phrase* to include the same information that was in the active-voice sentence and say: *The sample was filtered **by me**.* But a lot of times, the passive is used precisely because we don't want to emphasize who is performing the action—we want to emphasize what happens to the sample. This use of the passive voice is what you often see in the *Methods* sections of scientific papers.

The common use of the passive voice in scientific writing has been a controversial subject for some decades. Here are some of the arguments *against* use of the passive voice:

1. It is more complex, and thus might be more difficult to understand, especially for those people for whom English is a second language. Because so much science from all over the globe is now published and read in English, it should be in a simple,

easy-to-read style, because English won't be the first language of many readers.

2. It gives a false sense of objectivity to the writing; humans do science, and therefore the sentences describing how that science was done should include the humans. To some critics, leaving out the humans by using the passive voice is a way of pretending that science is not influenced by human biases.

On the other side, those arguing *for* the use of the passive voice point out the following in response to the above critique:

1. It isn't that hard to understand the passive voice; simple, clear sentences can be written in the passive voice.
2. In research projects with many people working together, naming individuals doing different parts of the research may make the research harder to follow and distract readers from concentrating on the things being studied. One could just always use the pronoun "we" in all cases, but that doesn't really change the information content of sentences.
3. Most scientists reading scientific papers are well aware of the problems of human bias in making observations and interpreting them; it's one of the first things that peer reviewers look for. Changing how sentences are written isn't going to suddenly cause people to become more aware of these issues.

Word processing applications have on occasion joined in the controversy on the side of those *against* the passive voice; some versions of word processing software will highlight sentences in the passive voice and ask you to consider revising them. You can turn this feature off, but I keep it on; often I do find ways of writing active voice sentences that are more direct, once the software has drawn my attention to them. On the other hand, I ignore the advice of the software when writing *Methods* or *Experimental Sections* of scientific papers. In such papers, the convention to use the passive voice is well established, and the writing in the passive when describing experimental work comes naturally to many scientists as a result of our exposure to this convention.

Different journals have different views on the passive voice; if you look through *Science,* you will find few sentences in the passive; *Science* editors encourage use of the active voice in their *Information for Authors.* On the other hand, a more traditional journal like the *Journal of the American Chemical Society (JACS)* will almost always have *Methods* or *Experimental Sections* written in the passive voice.

Given that there remains a lot of controversy about the use of the passive voice, it is best to be able to read and write well in both the passive and the active voice. If you send your paper off to *Science,* avoid using the passive as much as possible. If they reject it and you decide to send it to *JACS,* you will have to restructure it anyway (*Science* papers don't follow IMRAD format), and you might as well rewrite the *Methods* in the passive voice. So being able to switch sentences from active to passive, or passive to active, is a worthwhile skill to have; what is more, if you can do this, you will understand the basic grammar of your sentence better, and better understanding will help you simplify and clarify your writing.

There's one more *voice* issue you need to be aware of, and that is the *imperative voice.* This is the one where there is no subject; the sentence is asking you to be the subject, and to follow the directions: *Filter the sample. Rinse the precipitate. Dry the precipitate. Record the precipitate mass in your laboratory notebook.* Sound familiar? This is how laboratory instruction manuals are written. When you write proposals or papers, don't use the imperative voice. Save it for when you're writing instructions for students or fellow lab mates.

At the end of this chapter I have included a "Quick Guide" to tense and voice in scientific writing, showing how some simple sentences can be written in many different tenses and in both the active and passive voice.

12.2 GENERAL WRITING AND STYLE SUGGESTIONS

This book is not meant to be a complete guide to scientific writing; other texts, such as Angelika Hofmann's *Scientific Writing and Communication,*[1] deal with almost every conceivable grammatical and stylistic issue in scientific writing. Hofmann's book has plenty of helpful lessons and looks as if it would be particularly helpful to those for whom English is not their first language. I have found, however, that the following general suggestions cover the most common problems I see in student papers:

1. **Reread, and rewrite, your writing.** Back when I was in college, I still had a typewriter for typing papers, although computer-based word processing was starting to become more common. I'd write rough drafts of my papers with a pen on paper, and then type up a final draft. As I typed, I'd read the sentences before I typed them, and I think I completely rewrote about 50 percent of the sentences in the process of typing them. Sometimes I'd completely rewrite paragraphs. This really helped me learn to improve my writing.

   I remember one professor at our university being excited about the prospects for computer word processing. He thought that, because it would make rewriting so much easier, everyone would spend lots of time polishing up their writing, doing multiple revisions until little improvement was possible. The ease of editing would lead to better writing by everybody! I am a professor in a time when everybody uses word processing, and I'm afraid that this professor's prediction wasn't correct. Students *don't* end up critically reading their writing and imagining how to fix it. If a second draft is required, it is too tempting just to use what is already in the computer, with only a few superficial changes, and the awesome possibilities of word processing don't end up being used to improve writing.

   Be aware of the great opportunities for rewriting that word processing gives you. Read your writing over carefully and try to fix every sentence and paragraph that isn't perfectly clear and direct. Don't do this immediately after having written the first draft. Being away from the writing for even a few hours will cause you to look at it a bit more critically, a little more as a stranger would view it. Giving it a few days rather than a few hours is even better, but it means you have to finish the first draft a few days before the paper is due!

2. Identify the subject and the verb in troublesome sentences.
If, in reading your sentences, you find some that don't seem
to sound right, or don't seem clear, it is often helpful to start
analyzing what is wrong by finding the subject and verb. Write
(or just say in your head) a simple, three-to-five-word skeleton
sentence with subject/verb/object, getting rid of all the prepositional phrases, adjectives, adverbs, and other modifiers. Once
you have this basic skeleton, you can evaluate how to restructure the sentence. Then you can gradually add in all the details
while making sure you don't veer too much from the simple,
clear core of the new sentence.

To show how this process might work, I've taken the first
sentence from a *Results and Discussion* section in the chemical literature and changed a few of the nouns to protect the innocent:
"The crystal is firstly characterized by the UV-visible spectrum
and x-ray diffraction pattern shown in fig. 1."

There is nothing grammatically *wrong* with this sentence,
but it sounds awkward and isn't as simple as it could be. A first
guess at a skeleton sentence might then be "The crystal is characterized." This skeleton doesn't include an object; "is characterized" is the verb. Because the verb contains "is," a form of the
verb "to be," and the sentence contains a prepositional phrase
starting with "by," we can identify this as being in the passive
voice. With sentences in the passive, we should always check to
see if an active voice skeleton sentence helps clarify the sentence.
We can do this by making the nouns of the prepositional phrase
into the subject, and making "the crystal" the object: "A spectrum and diffraction pattern characterize the crystal."

Does this new sentence sound like a good sentence to you?
It doesn't to me, because spectra and diffraction patterns aren't
smart enough to characterize anything. *Scientists* must have
characterized the sample *using* the spectrum and diffraction
patterns. "We characterized the crystal using UV-visible spectroscopy and x-ray diffraction" is a more sensible active-voice
sentence. Notice that I changed it to the past tense, which is
more appropriate if we are talking about work the scientists
did. I also used "spectroscopy" and "diffraction" rather than

"spectrum" and "pattern"; the result is more direct. But this sentence is now more of a *Methods* sentence than a *Results* sentence, and it doesn't reference the figure, which might have been the *real* reason for the sentence in the first place!

If referencing the figure were the main goal of the sentence, perhaps "figure 1" should be the subject. Then the skeleton sentence might be "Figure 1 shows a spectrum and a diffraction pattern." We can then add in a modifying phrase to specify which spectrum and diffraction pattern are being shown: "Figure 1 shows the UV-visible spectra and diffraction pattern of the crystal."

This is shorter, clearer, and more direct than the original sentence. It is useful to consider what was left out? "Firstly" was left out, but since the original was the first sentence of the *Results and Discussion*, it's not really needed. "Characterized" was left out, but if the rest of the paragraph were about what the spectrum and diffraction pattern mean, it should be clear that characterization took place.

Do we really need a sentence just for pointing out what data is in a figure? If the paragraph that follows describes what the spectra and diffraction patterns mean, maybe we should just start with this interpretation, and refer to the figure along the way. "The UV-visible spectrum of the crystal, shown in figure 1a, has three peaks characteristic of Ni(II) in a square-planar coordination environment. The diffraction pattern of the crystal (figure 1b) indicates that the salt crystallizes in the *Pbca* space group...."

All that analysis and rewriting looks like a lot of work for just one sentence. However, all this analysis uncovered all sorts of possibilities for the sentence, including the possibility that it could be eliminated entirely! Just as with the process of coming up with creative research ideas (chapter 9), when writing up your work you sometimes have to explore a wide variety of alternatives before you figure out the clearest and most efficient way to get your ideas across. As you get more experience in rewriting, you will get better and faster at spotting problem sentences and finding ways to fix them. Finding the skeleton subject-verb-object of the sentence is a good place to start; you can then compare it to what you think the sentence should be about, and make revisions as needed.

3. **Check the length of your sentences.** Scientists typically use procedures that are lengthy and complex, and they are often trying to keep track of many different lines of data and observation at once. As a result, what is in their brains can be pretty complex. In trying to express what you did and what you found to other scientists, you may be tempted to try to express all this complexity in just a few sentences. This can lead to some really long compound sentences that are not best for communicating procedures or results to other scientists. Read your sentences back to yourself. If you find yourself running out of breath in a single sentence, see if there is a way you can break that chain of thought into two, three, or four separate, simple sentences. Identifying key subjects and verbs, as I illustrated in the previous suggestion, can be very helpful in deciding how to break up long compound sentences. If the subjects and verbs of one half of the sentence don't have much to do with those in the other half, put them in separate sentences.

Here's an example of what I might be tempted to write in a *Methods* section: "After the 24-well plate was prepared with solutions having different calcium concentrations in each well, it was placed in the crystal growth chamber, and a 50 mL beaker containing 15 mL of carbon-dioxide-saturated water was placed next to it so that carbon dioxide would gradually diffuse out of the water and into the solutions in the 24-well plate once the chamber was sealed."

Like the example in the previous section, there's nothing grammatically *wrong* about this sentence. It is a compound sentence where two complete, grammatical sentences are joined by the conjunction "and." But each of these sentences has multiple clauses, and one of these clauses itself employs "and."

To illustrate the range of possibilities open to us in rewriting this sentence, I will first go to the opposite extreme of writing a sentence for each idea: "The 24-well plate was prepared. Each well had a solution with a different calcium concentration. The plate was placed in the crystal growth chamber. A 50 mL beaker containing 15 mL of water was saturated with carbon dioxide. This beaker was added to the growth chamber. The beaker was placed next to the 24-well plate. The chamber was sealed. Carbon dioxide gradually diffused out of the water. Carbon dioxide diffused into the solutions in the 24-well plate."

That's not great, but, for scientific writing, it's actually *better*, because every sentence is a simple one subject-one verb sentence. All but the last two sentences are in the passive voice; the last two sentences are in the active voice, with "carbon dioxide" as the subject.

Splitting up each of the ideas into its own sentence, however, may not be the best way to go. Too many short sentences make your writing choppy and repetitive. Some of the sentences are closely related to each other, using the same words, and it would be good to put them together in the same sentence. Here is one attempt to strike a middle ground between the two extremes: "Solutions of different calcium concentrations were prepared in a 24-well plate, with each well having a different calcium concentration. The plate was placed in the crystal growth chamber. A 50 mL beaker containing 15 mL of water was saturated with carbon dioxide and added to the growth chamber, next to the 24-well plate. The chamber was sealed. Carbon dioxide gradually diffused out of the water and into the solutions in the 24-well plate."

4. **Don't try to express your personality or to be creative in your writing.** You want to express your *ideas* as well as possible, but not your individuality. Many instructors who teach writing outside the context of science may have urged you to "find your voice" and to give your writing a unique personal stamp, an expression of your personality. In science, however, this can just get in the way of other scientists' ability to judge your methods, results, and conclusions objectively. The extra humor, flair, or color in your writing may actually prejudice readers against the science you are trying to convey. Scientific writing should sound as if *any* scientist might have written it.

For scientists, the interesting part about a scientific paper is the data and interpretations being communicated, not the writing itself. Here's one way to think about this: a novelist is very often writing about the same things novelists have always written about—love, hope, sadness, joy. An exciting new novel is exciting because it explores these ancient topics in a new way, in new contexts, with new words and new ways of writing, so that the reader can experience these things in a new, fresh way. On the other hand, what the scientist is trying to describe is often so new, so previously unheard of, that there's no need to dress it up. You just need to calmly, logically convince people that it is actually true. Think about the discovery of atoms, relativity, quantum mechanics, DNA—all of these discoveries were just *unimaginable* to people living just a decade before they were announced to the world. Interesting new ways of writing were not necessary to announce these discoveries and make them exciting. What was needed were very calm, careful, logical arguments convincing people that these surprising discoveries were true. The last thing a scientist would want to do when announcing such exciting new results would be to give readers the impression that they are just being clever, funny, emotional, or grandiose.

5. **Be cautious.** Scientific writing is filled with cautious verbs, careful specifications, and qualifying clauses. In chapter 1 and chapter 7 we looked at how tentative new conclusions in science arise out of many experiments, and relative certainty only comes after many efforts to falsify or negate these conclusions fail. Any research that is new enough to be exciting is also new enough that any conclusions are tentative.

If a single set of experiments show that some soybean plants grew faster when the soil was inoculated with certain bacteria, you should not write, "This experiment proves that beneficial bacteria make plants grow faster." Rather, the limitations of the experiment should be kept in full view. A better summary of conclusions might be the following: "In this experiment, *Glycine max* plants benefitted from soil inoculation with *Bradyrhizobium japonicum* when grown in sandy soil. These results suggest that some interactions between host legumes and bacteria can be beneficial, but more research will be needed to determine whether the benefits of these interactions extend to other growing conditions and other species of legume."

Notice that in the second summary, I use the cautious verb "suggest" rather than the bold verb "prove"; I carefully specify what species were used and some of the growing conditions, as the benefits may be specific to those species or those growing conditions; and I acknowledge that more study will be needed. The scientists who read your writing know that there are always limitations to research projects, and that new conclusions are frequently tentative. If they can see in your writing that you know this, too, then they will be more likely to think of you as a careful and thoughtful scientist, and will be more likely to take your research seriously.

FOR FURTHER STUDY AND DISCUSSION

1. Go to the scientific literature and find at least one sentence each written in the past passive, present passive, future passive, past active, present active, and future active tense. (If you can find some in any of the perfect tenses, include those for bonus points!) Identify the subject, verb, and object (where present). Rewrite active sentences in the passive, and passive sentences in the active.

2. Your instructor will provide you with sample sentences from the literature, and you will be asked to identify the subject, verb, and object (where present) and to identify the voice and tense of these sentences. For sentences in the passive voice, rewrite in the active voice; for sentences in the active voice, rewrite in the passive voice.

3. Find an awkward, hard-to-read sentence in the scientific literature. Figure out what the skeleton sentence is, and come up with five different ways of rewriting the sentence or moving the information in the sentence to other nearby sentences.

ADDITIONAL READING

Hofmann, A. H. *Scientific Writing and Communication. Papers, Proposals, and Presentations* (Oxford University Press, 2016).
   This book is very comprehensive, including writing tips for those for whom English is not their first language.
Penrose, A. M. & Katz, S. B. *Writing in the Sciences. Exploring Conventions of Scientific Discourse* (Pearson Longman, 2010).
   This book is more focused on style and rhetorical issues than Hofmann's book and has good suggestions about writing cautiously.

A QUICK GUIDE TO TENSE AND VOICE

Some simple English grammar as it applies to writing scientific material.

PASSIVE VOICE

Past	The solid was filtered.
	With prepositional phrase: The solid was filtered by me.
Past Perfect	The solid had been filtered. (rarely used)
Present Perfect	The solid has been filtered. (rarely used)
Present	The solid is filtered.
Future	The solid will be filtered.
Future Perfect	The solid will have been filtered. (rarely used)

The **past tense passive** is most often used for the *Methods* or *Experimental Section* of scientific papers. The prepositional phrase indicating the actor is, in my experience, never used in this context.

The **present tense passive** is often used for laboratory manuals or instructions on how to do something. The present is used because the writing is not referring to one particular implementation of the procedure but to all implementations in the past, present, and future.

The **future tense passive** is suitable for proposal writing, because you are describing what you intend to do in the future.

ACTIVE VOICE

Past	I filtered the solid.
	The volcano sent tons of ash into the atmosphere.
Past Perfect	I had filtered the solid. (rarely used)
	The volcano had sent tons of ash into the atmosphere.
Present Perfect	I have filtered the solid. (rarely used)
	The volcano has sent tons of ash into the atmosphere.
Present	I filter the solid.
	The volcano sends tons of ash into the atmosphere.
Future	I will filter the solid.
	The volcano will send tons of ash into the atmosphere.
Future Perfect	I will have filtered the solid. (rarely used)
	The volcano will have sent tons of ash into the atmosphere.

For the active voice, I have included two kinds of sentences: one with a human subject, which gets at the heart of the controversies surrounding use of the passive and active voice, and another sentence that involves a non-human subject. Notice that, if you rewrote the second, non-human-subject sentence in the passive, you'd *have to* include the prepositional phrase to have all the information contained in the original active-voice sentence: "Tons of ash were sent into the atmosphere *by the volcano*." The volcano is too important to be left out. In this situation, you would be better off using the active voice, as it would result in a simpler, clearer sentence.

The active voice can almost always be used, although it is a common convention in many journals and disciplines *not* to use it in the *Methods* or *Experimental Section*. Some people used to insist that the first person (*I, we, me, us*) not be used in scientific writing, but use of the first person is now generally accepted, and some great historical papers have included use of first-person pronouns, as in the example below:

"We wish to suggest a structure for the salt of deoxyribose nucleic acid." Watson, J. D. & Crick, F. H. C. Molecular structure of nucleic acids: a structure for deoxyribose nucleic acid. *Nature* **171**, 737–738 (1953). Notice that this sentence is in the active voice and the present tense.

The present tense, active voice is often used to make general conclusions about the object of study: "The electronic absorption is due to a transition between two partially filled d orbitals." "The benzene reacts under high pressure." "Centrifugation separates the nicked and whole plasmids."

In these examples, the subjects, or actors, are the observed phenomena, the atoms or molecules, or the processes used, rather than the scientists, so no use of first-person pronouns is needed. Notice also that the present tense is used, because it is believed that these statements not only *were* true for one past experiment but *are* true now and *will be* true in the future.

On the other hand, if you are only referring to what happened in your particular experiment, the past tense is used: "Various solvents *contaminated* our sample, and, as a result, the NMR spectrum *was* complicated."

Many people will tell you never to mix tenses, but there are circumstances in which it can be appropriate. Imagine if the sentences above were immediately followed by general statements about the phenomena being studied:

"Various solvents **contaminated** our sample, and, as a result, the spectrum **was** complicated. Nevertheless, our results *demonstrate* that the metal-bound proton *can be observed* by NMR, and generally *has* a chemical shift upfield of the TMS standard."

Notice that I shifted here from the past tense (bold) when writing about our particular experiment that happened in the past, to the present tense (italics) when talking about the results that still are shown in our laboratory notebooks, and the protons and their general behavior in the past and future.

CHAPTER THIRTEEN

ASSEMBLING AND WRITING A SCIENTIFIC PAPER

LEARNING OBJECTIVES

After reading sections 13.1 to 13.6, you will be able to:

- distinguish between what belongs in the *Results* and what belongs in the *Discussion*
- explain the advantages and possible disadvantages of keeping the *Results* and *Discussion* separate
- identify whether it is best to display particular data in a graph or a table
- identify formatting choices that will make a graph or table conform to traditional graphical standards and make the graph or table easier to understand for the reader

After reading section 13.7, you will be able to:

- describe the most difficult challenges in writing a good *Methods* section
- identify common strategies for organizing a *Methods* section

After reading sections 13.8 and 13.9, you will be able to:

- identify common topics covered in *Discussion* sections
- identify strategies for determining what needs to be discussed in the *Discussion*
- explain the general organization of a *Discussion* section
- explain why *Conclusions* sections are sometimes a good idea

After reading section 13.10, you will be able to:

- explain the role of the *Introduction* in a scientific paper
- identify the main parts of an *Introduction* and its usual organization

After reading sections 13.11 and 13.12, you will be able to:

- identify the different roles abstracts have played in the history of scientific publication
- identify how abstracts vary depending on the discipline and journal
- explain the most important considerations in deciding on a title for a paper
- identify common grammatical forms for titles and explain the advantages and disadvantages of sentence titles

13.1 SOME PERSPECTIVE

There are lots of books on writing in the sciences. The bibliography for this chapter gives an annotated list of some that I have looked at and found useful. Rather than trying to rewrite what so many have written before, I am going to merely give you an outline of how I like to write scientific papers, especially when it varies from what others have written about writing. I'm calling this chapter "*Assembling* and Writing a Scientific Paper" because a lot of the effort in getting a paper ready is not in the writing but in getting the right figures, tables, references, and descriptions of methods and results together into a cohesive whole. Writing is indeed part of this; you need to make sure your words, phrases, and sentences smoothly guide the reader over the whole assembly. But at the beginning, it's more important to have the right pieces in the right places to begin with.

Back in chapter 4, I discussed how, in the twentieth century, it became common for science papers to be organized in a standard format consisting of *Introduction, Methods, Results*, and *Discussion*, or IMRAD. But I doubt if there was ever a time when *all* scientific papers were written in this format. By the time everyone had gotten the message that it was good to arrange papers in the IMRAD format, some people were already rebelling against the inflexibility of that format for some research projects, or insisting that the *conventions* in their own discipline were more important than the IMRAD format. In the magazine-like journals *Science* and *Nature*, papers don't have the labeled sections—although the same sort of information is typically contained somewhere in these articles and reports. Especially confusing for some readers of these journals is that the methods are typically described in figure captions and footnotes. In more traditional journals, *letters* and *communications* typically do not follow the IMRAD format, even if, once again, the information is there somewhere.

So, before we get into the details of what goes where in a scientific paper, we have to recognize that the IMRAD format is more a *convention* than a *rule*. Editors will often strongly encourage authors to follow this format, but if a paper still doesn't conform to the format after peer review, editorial suggestions, and revision, it still has a good chance of being published if peer reviewers think the work deserves to be read by other scientists and added to the archive of scientific knowledge.

13.2 AUTHORSHIP

It is important to realize that *most* of the papers published in the scientific literature have more than one author. In this chapter I describe how you would set out to write a paper if you were the only author, because that's what students in my undergraduate class do. If you go on to publish science beyond the undergraduate level, however, you will likely be only one of many authors. Publications from an academic laboratory may have two graduate students, a postdoctoral fellow, and a professor listed as coauthors; a paper from government or industry may also have half a dozen colleagues and their supervisor listed as coauthors. As science becomes more complex and interdisciplinary, more people from multiple laboratories or institutions get involved. In high-energy physics, discoveries of new particles require the design and construction of huge particle colliders and detectors, and the papers reporting these discoveries may list hundreds of scientists and engineers as coauthors, each involved in some aspect of the design and testing of the apparatus, as well as in the data collection and analysis that ultimately led to the discovery.

Who actually gets to be an author, and where on the list of authors they are placed, can become something of a matter of concern and even controversy. Different journals and institutions will have slightly different guidelines, but most agree that, to be an author, you need to have *contributed substantially* to the work and *share responsibility* for what gets published. If you are listed as an author on the paper, and someone asks you about the paper, you had better be able to say something more about the paper and its reliability than, "I don't know, I just plated the bacteria and counted the colonies." In other words, technicians shouldn't be authors—they can just be acknowledged in the *Acknowledgments* at the end of the paper. At the other end of the spectrum, sometimes people who are high up in institutions get added as authors in honor of their positions of importance even if they contributed neither much work nor thought to the paper. Many people and journals agree that these people really shouldn't be authors; however, if they are, they need to be willing to take responsibility if the data or conclusions in the paper turn out to be mistaken or even fraudulent.

Adding another layer of complexity to the issues of authorship is the order in which the authors are listed. In some disciplines, the head of the laboratory—the professor, in the case of an academic laboratory—is listed last, and the graduate student for whom the work was their principal project is listed first. Other involved graduate students and postdoctoral fellows are somewhere in the middle. In other disciplines, the head of the laboratory is first, and everyone else follows. Usually, one or two authors will get an asterisk or other mark next to their names indicating them as "corresponding authors." This is the author you would write to if you had questions about the paper, and it's usually the head of the laboratory. Some journals have now started requesting that the contributions of individual authors be described in a footnote, making it easier for a reader to figure out which authors were responsible for different aspects of the work.

Because all authors must share responsibility for a published work, all authors should contribute to the writing and/or editing of a paper. This is going to work differently in different laboratories, but ultimately everyone needs to have seen the final draft of the paper and agreed with all of it. Often different individuals who did different experiments are asked to write up corresponding parts of the *Methods* and *Results* for the paper, while the head of the laboratory writes the *Introduction* and *Discussion*. But ultimately you need to know how to write *any* part of the paper, because once the final paper is put together, you will share responsibility for all parts. You may end up having to explain to the head of the laboratory why the *Introduction* or *Discussion* needs to be changed; it is best to do that diplomatically so that a suitable compromise can be reached!

13.3 STARTING WITH THE RESULTS

Before you begin to assemble a scientific paper, you do experiments or take observations. Your ideas for those experiments may have been influenced by literature you read, and many references to this literature will occur in the *Introduction*. You may, therefore, think that the most natural thing would be to start by writing the *Introduction*. While some scientists may do this, it's not where many experienced scientists start. It's often better to write the rest of the paper first and then design your *Introduction* so that it sets the stage for—introduces—the paper you have assembled.

In the past I had students start by writing the *Methods*. This is usually the easiest thing to write, because you're just describing what you did. Thinking back over published papers that I coauthored, however, I realized that papers often start by considering what results there are that make writing a paper worthwhile. There's no point in starting your paper writing about experiments whose results you may decide aren't worth including in your paper. It is best to first decide what the key results are, as these will determine which methods you will include.

In an undergraduate course, including class laboratories and ones where you do a simple at-home research project, there probably aren't too many experiments; you're going to have to write about the few experiments you did, regardless of the results you got. But it is reasonable to look at your data, decide what is important, and ask yourself what kind of story you are going to be telling with your paper. Sometimes the results end up answering a different question than the one you originally asked. It makes sense to think about the results you have and how you will interpret them in the *Discussion* before writing anything. In academic research laboratories, this part of assembling a paper often occurs in conversations within a lab group; the discussion will frequently be about answering the question "Do we have a paper here?" In other words, "Are there results that people will be interested in, provided we help them interpret those results in our *Discussion?*"

13.4 DISTINGUISHING THE RESULTS AND DISCUSSION

In my experience, one of the things that is hardest for students to do in writing a paper is to clearly and correctly set the *Results* apart from the *Discussion*. (I didn't do a good job on this in the first draft of a paper I wrote as a graduate student—but my research advisor quickly set me straight!) The traditional IMRAD format clearly separates the *Results* and *Discussion*. But a look at scientific journals in a variety of fields will show you that this is one of the conventions that is followed least often, as there are many papers with a combined *Results and Discussion* section. In some cases, melding the *Results* and *Discussion* together may be appropriate. But I believe that, for science students, it is good practice first to try as hard as you can to clearly separate what belongs in the *Results*, and what belongs in the *Discussion*. If doing this makes your paper hard to read or understand, then you may be justified in mixing the two. But first make the effort to separate them.

In chapter 4, we discussed one reason the IMRAD format came about; busy readers could benefit from having a standard format, so they could reliably find the information they were looking for. But I think there is something more fundamental about the convention of separating the *Results* and the *Discussion*. Presenting the *Results* separately from the *Discussion* helps the reader more objectively evaluate the work. It encourages the reader to first encounter the data and observations without the authors' interpretations. The reader has a chance to think for themselves about what the data might mean and draw their own conclusions. Then they can read what the authors think their results might mean in the *Discussion*, and compare their interpretations with those of the authors. Putting the *Results* off in a separate section encourages both the reader and the writer to clearly acknowledge that there is a difference between the facts of the *Results* and the interpretations of the *Discussion*.[1]

I advocate clearly separating the *Results* and *Discussion* because writers, especially student writers, need to think about what might be regarded as a straightforward, easily accepted fact by the scientific community and what might be considered an interpretation that other scientists may disagree with. When we get to oral and poster presentations of research, in chapter 14, you will see that I don't recommend separating the presentation of results from discussion of those results; in giving a talk, experiments, their results, and their interpretation can be used to tell a story that keeps an audience awake and following along.

13.5 RESULTS, SELECTED AND PRESENTED

In most real research projects, you will have much more raw data in your notebook than you can present in a paper. The data you present will have to be *selected* from all this data, and it will have to be *presented* in a way that favors the interpretation you give in the *Discussion*. When I say the data is *selected* and *presented*, however, I don't mean that contradictory evidence is ignored or that it is presented in a deceptive or dishonest way. That would be unethical! The *selection* of data may be a matter of deciding not to mention all the different experiments that gave you no information because they were poorly designed or rested on false assumptions. The *presentation* of data may involve deciding whether to present every trial separately or to average all the different trials and show standard deviations. It also involves questions of whether to present numerical data in a table or a graph.

Selection. When we examined how to read a scientific paper, one of the first things we recommended was to look at the pictures—the graphs, tables, and original data—that presented what was being done and what was found. It is also good to start with the pictures when assembling a scientific paper. If you've been keeping a good laboratory notebook, you may have already started on the pictures. You already have tables of data, and, if you've taken the *Discussion of Results* and *Conclusions* sections of your notebook entries seriously, you may have already graphed these results to see what trends emerged. But the graphs and tables that are in your notebook may not be the graphs that end up in your paper. You need to decide what the most important aspects of your data are and how you will most effectively communicate them to the reader.

One of the decisions that typically has to be made with quantitative data is whether to present it in a graph, in a table, or just in the text. With journal space at a premium, most reviewers and editors won't allow you to do both a table and a graph of the same data. Graphs are more useful when you are trying to show trends or how the data conform to mathematical functions. They are also useful when they represent raw data in a form that everyone will quickly recognize as an NMR spectrum, a current-versus-voltage curve, or an electrocardiogram.

If the actual numbers themselves are important, tables are better. For example, if you are showing values of ΔG°_{f} (the free energy of formation) that you measured for a series of compounds, it would be better to present them in a table, because these are numbers that people in the future will probably want to use for calculations. Also, if there isn't a clear trend or mathematical function for a collection of numbers, a table may be better. One of the questions I ask myself is whether the x-axis of a graph *means* anything, whether it makes sense as a quantitative measure. For example, if one is comparing different laundry detergents, one could make a column graph (sometimes called a bar graph) for the different detergents, but there's no particular reason to arrange them in any one order on the x-axis. It's not a quantitative independent variable; a table might be more appropriate.

With every graph or table, you should ask yourself, "Does it really show something that I can't just convey in words?" If your data fits well to a mathematical function, it may be worth it to show the reader just how good the fit is on a graph. On the other hand, if you just have four data points that show that the sugar content of a tomato *generally* increases with ripeness, and you can't show a clear mathematical dependence or even a smooth trend, it might be better not to use a graph, but just to describe the trend in the text of the *Results* section. Always remember that journal space is scarce, and you shouldn't use it up with tables or graphs of questionable value.

Another decision you will have to make is how much to reduce quantitative data through averaging or selection, and how clearly you want to indicate the spread of the data. This is where you have to be really careful about being honest. If you did a number of trials of exactly the same experiment and got a small spread in the numerical data, it makes a lot of sense just to present averages and standard deviations, or standard error bars on a graph. But what about experiments where you slowly worked the "bugs" out of the experiments and you have the most confidence in the last one? Well, the ideal thing to do is repeat the experiment several times, exactly as you did it the last and best time, so you can present proper statistics. However, there will be cases where this is prohibitively expensive or time-consuming. Then the best thing to do is to present the data and explain, honestly, in the *Results* text something that clarifies the situation. For example: "Data shown in table 1 is from the last of three trials. Although the general trends were the same for previous experiments, the yields were somewhat lower. We believe that this is due to less scrupulous attention to anhydrous conditions in the earlier experiments." Ideally, if you have room, you would also give the full data for the earlier experiments—but the editor and peer reviewers may ask you to remove them to save space.

There is one bit of slight dishonesty in the presentation of data that is pretty common in the scientific literature. It's even a joke among scientists, because all experienced scientists are aware of the issue. This is the situation where there is a spectrum, a photomicrograph, a picture of a stained electrophoresis gel, a voltage-versus-time trace, or some other display of raw data. The text will say "a *typical* spectrum (picture, photomicrograph, etc.) is shown in figure 2," or perhaps "a *representative* spectrum." Everybody knows that "typical" or "representative" probably means "the best example we had." If typical data really isn't anywhere nearly that good, and there's no good explanation for why, then such a presentation of data is unethical. But if there are mostly *superficial* reasons why other spectra, photomicrographs, or gels don't look quite as good, it's probably okay. Often there may be data that is good enough to convince you of a certain conclusion, but your research advisor will want you to do it one more time, paying attention to all the details so that you get a "prettier" gel picture or spectrum for publication. You may want to discuss this ethical issue with your fellow students, instructor, or research advisor!

Presentation: Graphs. Before there was graphing software, there were graphic artists at work in most science departments of major universities. If you wanted to publish a graph of some data, you'd take the data to the graphic artist. The graphic artist knew all the rules and conventions of good graphical presentation, because they were trained in them. Toward the end of the twentieth century, computers and software that could make graphs became cheap and common. Suddenly, everyone was their own graphic artist—but most had not been trained in the graphic arts and didn't know how to make clear, easily readable graphs following the usual conventions. Editors of scientific journals during this period often wrote pleas in their annual *Information for Authors*, trying to educate potential authors on graphic standards.[2] More recently, however, it appears they have largely given up on giving much specific advice. Perhaps this was because authors have become more sophisticated in their graphic skills; perhaps editors have given up instructing people in advance, leaving the correction of poor graphics to the peer review process; or perhaps editors have just given up on what they see as a hopeless cause.

It is true that some of the old graphical standards have become obsolete. For a long time, and even in some recently published books about how to write science papers, it was a rule that everything needed to be in black and white for ease of printing. Many journals started accepting color graphics in the 1990s, but only if the author was willing to pay any increase in the cost of printing, and the editors could be convinced that the color was essential to the clarity and informational value of the figure. Biology journals, where color photographs and micrographs were essential to conveying observations, started accepting color figures a lot earlier than chemistry. But as journals are increasingly read online rather than in printed form, the economic barriers to the use of color have crumbled, and many important journals now routinely accept color graphics without charge to the author.

Nevertheless, it's still worthwhile to pay attention to the old graphical standards, and to practice making clear graphics in black and white. For one thing, there may be plenty of situations in which color is *not* freely available. Making black and white photocopies for distribution of reports and papers is still cheaper than making color copies. Color is perceived differently by different people (especially in the case of color blindness), and colors that do not show clear contrast can make graphics tiring to decipher. The discipline of making a good black-and-white figure is helpful in thinking about how to make the best possible color figure.

While computer applications have made it possible for every scientist to be their own graphic artist, not all applications can easily generate graphs that represent best practices in making graphs. In my experience, the most popular spreadsheet programs that students are familiar with aren't really set up to do publication-quality graphs. Because the software market is constantly changing, and because I've never found the "perfect" product, I'm not going to make any particular recommendations, but graphing software that

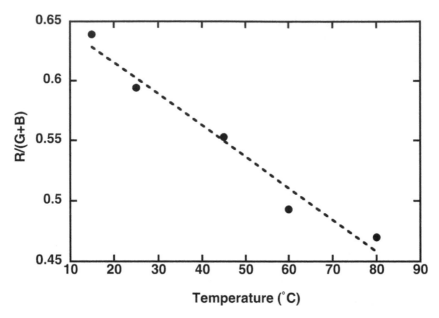

Figure 13.1. A plot of the ratio of average red (R) values to the sum of average green (G) plus blue (B) values for 24 x 24 pixel samples of digital photographs of lipstick stains on a white t-shirt, after washing with Nuke-It brand laundry detergent, as a function of wash temperature. The data is fit with the linear function $R/(G+B) = -0.0026\ T + 0.67$.

is more focused on making publication-quality graphs is available, and it often isn't too expensive.

Take a look at figure 13.1. It's a pretty decent graph, made with software designed to make publication-quality graphs. The vertical axis is a bit hard to understand, and doesn't have units (because ratios don't typically have units!), but the formatting makes for a clear picture of the data. Figure 13.2, on the other hand, is the same data graphed from a common spreadsheet program, with few fixes to the default graph settings. The same spreadsheet software *can* be made to do better, but my students who have really worked at it tell me it's hard.

Here are some of the rules that you should keep in mind when making a graph of data. Look at figures 13.1 and 13.2 to better understand how they are being followed, or not followed, in those graphs.

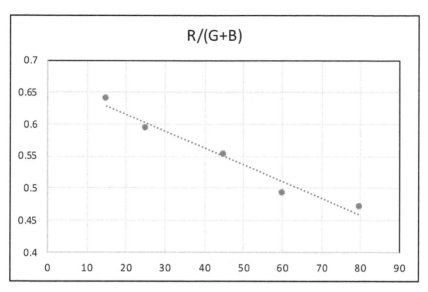

Figure 13.2. The same data as plotted in figure 13.1, but with poor formatting of the graph.

1. Don't have a colored or shaded background for the graph. White may be boring, but it provides maximum contrast, and better contrast makes for easier reading.
2. Don't have gridlines within the graph. It makes the graph too "busy," and there can be problems maintaining contrast and visibility without dominating the data itself. You may think that it would be good to have gridlines so that readers can better estimate the numerical values of the data, but this is not what a graph is for. A graph is for illustrating trends in data; if the actual numerical values of particular points are important, they're better off in a table.
3. When possible, don't just show the x- and y-axis at the bottom and left-hand side of the graph. Instead, have lines on all four sides of your graph, making a nice box around your data. A lot of applications don't readily do this for you, and, based on what I see in published journals, a lot of journals don't insist on it, but it's part of the traditional graphical standards.
4. It is customary to have *tick marks* to show where some numerical values fall on the x- and y-axes. Because you're just trying

to show trends, and not give a source of exact numerical values, these tick marks don't need to be too closely spaced. I was taught to always have tick marks *inside* the frame of the graph only, all around all four sides. But some applications refuse to do this, no matter how hard you try, and this is one of the standards that isn't carefully followed.

5. I shouldn't need to remind you that the *axes* of the graph should be labeled, and that the numerical values should also be clearly labeled. Maximum and minimum values on the axes need to be "appropriate," but there is some controversy about what this means. Sometimes people insist that zero always has to be shown on the y-axis, but in different situations, this is neither practicable nor appropriate. Think instead about the message you want your graph to convey, and how best to convey it, without being dishonest or misleading. In figure 13.1, for example, using zero as the minimum value for the $R/(G+B)$ ratio doesn't make sense; that would indicate that the white t-shirt ought to turn a blue, green, or bluish-green color after washing! A white or neutral gray color, where $R=G=B$, would have a $R/(G+B)$ ratio of 0.5.

6. Consider how the graph will look when it is in its final published size. Text that is readable when the graph is printed out at 6×6 inches or 15×15 cm may become unreadable when reduced to fit in a single printed column in a journal. The same applies to data points, and especially to the thickness of axes and lines on graphs. This can especially be a problem if you try to directly use raw data from a scientific instrument. The software of a nuclear magnetic resonance spectrometer may try to make the lines on an 8.5×11 inch or A4 printout of a spectrum very fine to make the resolution of fine structure more visible, but if these same spectra are shrunk to smaller dimensions for printing, the lines may not be visible!

7. Keep the font size and typeface consistent for all axis labels. Some journals specify sans-serif fonts for these.

8. If you have more than one data set on a graph, make sure that it is easy to distinguish the different data sets. Typically, this is done with different symbols for the data points—a triangle for

one, a circle for another, and a square for the third. Make sure, however, that the sizes of the data point markers are all about the same and are easily distinguishable.

9. Many graphing applications have an option where the individual data points can be connected by line segments or smooth curves that go *through* every data point. In general, don't use these; they may only be useful when you're trying to plot a spectrum or time course from numerical data. (In those cases, you are emulating an old-fashioned chart recorder, in which the pen is constantly in contact with the paper.) But if you have discrete data points separated by a fair amount of distance on the *x*-axis, the only lines that should appear are lines that represent mathematical functions that are *fit* to the data. In your earlier science classes, you have probably done a fair number of *linear* fits, where linear functions of *x* are fit to the *y* data points. But many programs also allow polynomial, exponential, logarithmic, and other mathematical functions to be fit to the data. Sometimes the points will all lie close to the line and at other times they will not, but the reader can judge for themselves just how good the fit is. If you can fit your data to a function, however, you should include in your paper why this function makes sense. Most often, functions will be used because they fit a particular *model* developed to predict how a physical or living system will behave.

10. Many applications will automatically put a title above all graphs. Remove this; when published in a journal, your graphic instead will have a caption *below* each graph, fully explaining what the graph shows. It won't say "Graph 1," though; all graphs are labeled as *figures*, as are all the other graphics and pictures in the paper. The caption should explain the figure well enough so that somebody just trying to get a sense of the data you collected won't have to refer to the text to understand the figure.

Presentation: Tables. As with the graphs, we have to remember that journal space is scarce, and we have to consider just how important a table is for presenting our data. If you think you can convey all there is to get out of a table in a few short sentences, you should consider whether or not you need the table.

A lot of people are used to cranking out tables for class lab reports using readily available spreadsheet applications. Journals, however, have their own standards for tables, and they often prefer that tables be formatted in word-processing programs. The American Chemical Society prefers that tables be formatted using the table feature of Microsoft Word or WordPerfect, rather than just using the tab key. Here are some conventions for making tables that are appropriate for publication:

1. The arrangement of the data is generally the opposite of what you might expect for a bar graph of the same data. The different samples go *down the columns*, and the individual data values for each sample are spread *across the rows*. For example, if you were tabulating the recommended wash temperature, the pH, and the effectiveness of various laundry detergents, the brands of laundry detergents would be down the first column (the *stub column* or *entry column*) and the temperature, pH, and effectiveness would be spread across the rows (see table 13.1). The variables that are more likely to be the independent variables are listed first (to the left), and those that are most likely to be dependent variables are listed last (to the right). In table 13.1, for example, we might expect the effectiveness of a detergent to depend on its recommended wash temperature, so we might consider the wash temperature as an independent variable, a variable that we control. If there really was a good correlation between these variables, a graph might be more appropriate, but in this case, no clear trend is evident, so a table is fine.
2. Horizontal rules (lines) are used to set the title apart from the column headings, to set the column headings apart from the actual data, and to set the footnotes apart from the data. The table feature of some word-processing programs will do this for you.
3. The title of a table should be at the top of the table—just the opposite of figure captions, which go below the figures. The title should clearly state what data is included in the table, so that the

reader doesn't have to refer to the text to understand what data you are presenting.

4. The column headings should present what data occurs in that column, along with any units that are uniformly applied to all of the entries in that column. The units are put in parentheses and often are shown in the second line of the heading.

5. Footnotes can be used to further explain the title, column headings, or even individual data entries. They can also be used to reference other publications, if data from those publications is being compared to data presented in the new study. Sometimes these references will not be direct; the footnote will just say something like "From ref. 1." Format will vary from journal to journal; in table 13.1 I have used the American Chemical Society convention, which uses italic lower-case alphabetic superscripts.

Presentation: Other Data. Images are also often used to display data. These include photographs of electrophoresis gels or organisms used in a study; micrographs; or processed images that reveal spatial dependencies of composition, temperature, or other variables. I can't list all the possibilities here; there will always be some new method of doing science that will yield other kinds of images that are now hard to imagine. The most important question to ask yourself is the same question you asked yourself about your graphs and tables: How

Table 13.1. The recommended wash temperature, pH, and residual lipstick stain intensity after washing with the detergents at the recommended temperatures.

Detergent Brand	Temperature (°C)	pH	Residual Intensity (scale value)[a]
E-Z Kleen	40	7.8	2
Nuke-It	75	6.5–7.5[b]	1
Hi-White	60	7.9	3
Ecorama	25	7.0	2

[a]The residual intensity was measured by comparison with a standard color scale, as described in the text. Higher values correspond to higher residual stain intensity, with 10 = original stain intensity.
[b]For this detergent, the pH value was heavily dependent on the pH of the original water to which the detergent was added, suggesting a low buffering capacity for the detergent.

much better than a text explanation is this image for conveying my results to the reader? Once again, if you can describe in a few sentences what an image shows, it may not be worth it to use the image. If you're not sure, submit the image and let the reviewers or editor tell you to take it out!

13.6 WRITING ABOUT THE RESULTS

A *Results* section can't just have tables, graphs, and other figures. There has to be some text, too. For a lot of laboratory classes, just presenting a table of raw data, or a graph of that data, is enough. But a good paper has to have both text and graphics.

You have to refer to the tables and figures in the text. I always like to include the reference to the tables and graphs as part of a sentence about the results, rather than just putting it in parentheses after the end of a sentence, although both are acceptable. In writing about the results, the best thing to do is to give the reasons why you chose to present the data in the table or graph. For example, in writing about the results in figure 13.1 and table 13.1, we might write the following:

> All of the detergents tested cleaned the lipstick stains off the white t-shirt to some extent, as shown in table 13.1. The highest stain intensity after washing was 3, which is considerably lower than the value of 10 for the lipstick stains before washing. There were differences in the cleaning ability of the different detergents, but these do not appear to correlate well with either the pH of the detergent in water or the recommended wash temperature. However, it should be noted that the most effective detergent, Nuke-It, also had the highest recommended wash temperature.
>
> For the Nuke-It detergent, there was a strong, nearly linear correlation between wash temperature used and the redness of the residual stain, as shown in figure 13.1. The fact that the final values of R/(G+B) are below the theoretical value of 0.5 for white or neutral gray probably reflects

the lighting conditions under which the photograph was taken. Cleaning effectiveness at different wash temperatures did not vary much for the other detergents, and there were no clear trends with temperature.

The first paragraph mentions the reasons why the data in table 13.1 was in a table rather than in a graph; there was no clear correlation with temperature or pH. But it also adds other information, as it puts the numerical values in perspective relative to the color of the stains before washing. But notice that we did not repeat all of the information shown in the table; otherwise there would be no point in showing the table. The advantage of having the information in the table is that it makes it very easy to compare the attributes and effectiveness of various detergents without having to read a paragraph full of numbers.

The second paragraph points out why the graph was displayed in figure 13.1; there was a clear correlation, and we wanted readers to see how close to linear this trend was. It may have been tempting to suggest that this is why Nuke-It has such a high recommended wash temperature, but that would be interpretation of the results, which we must save for the next section. Note that the last sentence discusses results that were *not* shown in either a table or a graph. This is fine; it's not worth the space to show graphs that didn't show clear trends, but it's important to mention what was found.

If you're just describing a few simple experiments, there is no need to divide up the *Results* into subsections. But if you look at journals, you will often see that more elaborate research projects will present different kinds of results under different subheadings. There may be a section for *Kinetic studies*, for *RNA flow cytometry*, or for *Materials characterization*. Whether or not you have these subheadings is up to you, but remember that readers will often want to zero in on a particular bit of information as quickly as possible. Very often, figures and tables in a printed journal are not immediately adjacent to the text that refers to them, so if a reader wants to read more about the kinetic data presented in, say, a hypothetical "figure 5," it might be good to have a clearly labeled *Kinetic studies* section in the *Results*.

13.7 METHODS

The *Methods* or *Experimental Section* is a good section to write after the *Results* have been written. Once again, the order in which we write an article reflects the order in which a scientist might "read" a scientific article. If a reader looks at the pictures (tables and figures) in the *Results* first, as I suggested in chapter 7, then that reader might very well have questions about how these results were obtained. Table 13.1 and figure 13.1 give a pretty good idea of what was found about the various detergents, but a reader may wonder about what color scale we used to get the "scale values" in table 13.1, or how the pH values were measured. The caption for figure 13.1 explains that the R/(G+B) ratio is calculated from a digital photograph, but, the reader might ask, what measures were taken to ensure that lighting conditions were always the same, and was the camera operated in a manual rather than an automatic mode? What software was used to determine the RGB values?

Thinking about the sorts of questions a reader might ask will help you focus on writing about the right things in the *Methods*. The most difficult part about writing a *Methods* section is knowing how much detail to include. The oft-stated standard is that there should be enough information for someone else in the field to be able to reproduce your work. But this rule is deceptively simple; not everyone who will want to reproduce your work will have the same background and experience. Inevitably, some reviewers and readers will think that you are providing too much boring detail, while others will be angry with you for leaving out what they think are important points. Knowing your audience, or at least guessing at your audience, is one of the more difficult and important aspects of scientific writing.

Typically, in writing for scientific journals, you can assume that people reading the journals know all about the standard methods in your field. Methods, however, constantly change, and what is a new method this year may be a standard method in another few years. The first paper that uses a new method will necessarily have to have an abundance of information in the *Methods*, so that other scientists can repeat the experiments. Most of the papers that follow this first paper will cite the first paper as a reference but may continue to include some details, especially if they have made modifications to the method. Once the method is fairly well established, there may be but a single citation in the *Methods* section: "Poly(dA-dT) was synthesized in a primed reaction with *Escherichia coli* DNA polymerase I (Radding *et al.*, 1962)." In the case of methods that almost everyone in a field uses, there may not even be references. For example, I have never seen chemical synthesis articles refer back to E. Büchner's 1888 paper in the *Chemiker-Zeitung* that first describes the Büchner funnel, which is often used in the vacuum filtration of synthesis products.[3]

As with the *Results* section, you may choose to organize your *Methods* around the different kinds of experiments you did, depending on the number and complexity of the methods used. If your paper is cited in the future as a source of a method, this can be especially helpful, because readers may be looking at your paper mostly to find out how a particular molecule was synthesized or a particular analysis was done. In organic chemistry, there is a set of conventions specifically designed to make it easy to find particular information for the synthesis methods and the product characterization.

If a new method described in the *Methods* section involved construction of a new apparatus, an explanatory figure showing a schematic or plan of the new apparatus might be necessary. Likewise, if experiments involved complex sequences of different procedures to construct something like a new gene or labeled protein, it is helpful to include an explanatory figure that helps the reader see the flow of the experiment as a whole.

13.8 DISCUSSION

The *Discussion* can be the hardest section to write, but it can also be the most exciting. The biggest challenge in writing a good *Discussion* may be to challenge your own assumptions about what your experiment means. If you are quite convinced that it means one particular thing, you need to consider what alternative interpretations of your data are possible. For many of my students who do small at-home science projects, a related problem often occurs; they think their results don't mean anything, because they didn't get the results they expected. Then the challenge is to figure out what the results *do* mean. The actual process of writing the discussion can help you with this. Sometimes, just putting your thoughts on paper and making sure that they form logical sentences will help you to see the flaws in your logic—or alternative ways of making your argument. Sometimes it will make you realize that your results are really answering a rather different question than the one that you originally thought you were asking. That is why I suggest you write the *Discussion* before you write the *Introduction*—you may not decide what your paper is about until after you write the *Discussion*.

One way to challenge your own assumptions is to *imagine* someone else looking over your shoulder and discussing your research with you. After all, a "discussion" in everyday life usually involves more than one person. The person looking over your shoulder ought to be someone who makes you a little uncomfortable with his or her skeptical, critical questions, someone who challenges your every assertion. "Are you sure? How do you know that? Does that necessarily follow from the evidence?" When your paper makes it out into the world, there will be people reading it who are at least as skeptical and critical as any reader you might imagine—although that depends on your imagination. The *ethos* of science (chapter 2) has skepticism as one of its principal norms, and the *organized skepticism* of peer review will happen to your paper before it ever makes it into print. So, it's a good idea to anticipate skepticism and criticism.

The overall direction of the discussion is usually from the specific to the general. Because the *Results* immediately precede the *Discussion*, the beginning of the *Discussion* is a good place to focus on the practical aspects of the data presented in the *Results*. What questions do you think people will have about your results? How good is the data? Emphasize why you think your methods were adequate and your data of sufficient quality. What possible sources of error are there in your method? How did you guard against those errors? If your results don't agree with what other people have claimed in similar experiments, how might you account for the differences?

After you have answered your imaginary critic's questions about the data you presented, you need to face the same questions you faced back when you were writing the proposal for your research: "So what? Who cares?" You have to explain to the reader why your results are important and meaningful. If you end up concluding that the data answers a completely different question than the one you originally asked, that's okay; just explain what you did learn from the data. You aren't obligated to tell the reader what question you had originally started out investigating. (Of course, if your experiment was for a laboratory class and your data doesn't really answer the particular questions you were *supposed* to be investigating, you may end up having to write a discussion that mostly just discusses the reasons why your data didn't result in answers to the original questions.)

The "So what? Who cares?" questions can be answered at a variety of levels, and it is best to explore as many levels of this as possible. Let's look back at the data we presented in the figures and tables in this chapter. Maybe we originally proposed answering the question of what laundry detergent is most effective. Table 13.1 shows that we could conclude that Nuke-It might be the best. That's what the data means on the most practical level. But table 13.1 also suggests that this conclusion has limitations; the high recommended wash temperature for Nuke-It means we can't use this detergent on items that can't be washed at high temperatures. This also needs to be in the *Discussion*.

It may have been that the data in figure 13.1 was obtained because we were curious about just how important the high wash temperature was—could it be that we don't need to use such a high temperature? Well, the data shows that this particular detergent gives a strong correlation between wash temperature and effectiveness, and this suggests that we may not want to use Nuke-It for cold-wash-only items.[4]

We can also go beyond these *practical* conclusions and start to consider some *theoretical* conclusions. Is the effectiveness of laundry detergents a function of temperature? Figure 13.1 suggests that this is the case for Nuke-It. But table 13.1 suggests that it's *not* just the wash temperature that determines the effectiveness of the different detergents, because Ecorama works better at 25°C than Hi-White does at 60°C. How can we explain that? Well, it's probably a matter of the different detergents having different compositions. Maybe different components of the different detergents work better at different temperatures. We could investigate the composition of the different detergents, test the effectiveness of individual components on their own, and maybe study the temperature-dependence of this effectiveness, but all that may be beyond the scope of the preliminary study that this article is about.

Which brings us to another topic often included in the *Discussion*: future work. Any good research project is going to provide at least as many questions as it does answers. If you held off publishing your work until all the remaining questions were resolved, you'd never publish! By discussing the next logical steps in answering these questions, you will be helping to reassure readers (including your imaginary and not-so-imaginary skeptics and critics) that you are a thoughtful scientist and understand the implications of your research. Often, scientists will not only portray these experiments as *possible* experiments, but experiments that are already *in progress*. Sometimes this is to warn other scientists not to put a lot of effort into what is already being done in the laboratory of the authors.

I hope that the questions and answers outlined above for the fake laundry detergent data give you some idea of the range of questions and answers you can discuss in the *Discussion*. As you did with your original research ideas in chapter 9, try to think *divergently* about all the possible meanings and interpretations of your data. If you were originally only focused on a practical question, try discussing your findings on a more theoretical level; if you had a theoretical question, don't avoid addressing practical implications. It may be that thinking too divergently will lead you to wild speculations that are unsupported by your data, and you may have to enlist your imaginary skeptic later to keep these out of the *Discussion*. Nevertheless, thinking through these possibilities will reduce the chances that a reviewer will surprise you with an alternative interpretation or flaw in your logic that you hadn't considered.

13.9 HOW ABOUT A CONCLUSIONS SECTION?

Although it's not a standard part of the IMRAD format, a *Conclusions* section is often added after the *Discussion*. This is especially useful if the *Discussion* is long and involved, with lots of careful argument about interpreting your results. The *Discussion* should be long enough to address the potential concerns readers may have with your interpretations of the data. But there will be other readers who just want to find out whether you've concluded anything that is of interest to them before plunging into all the detail of a lengthy *Discussion*. For this reason, a *Conclusions* section can be a good thing to have. The *Conclusions* is also a good place to state the conclusion of the *local argument* in your paper (see chapter 7).

13.10 INTRODUCTION

Having assembled the *Results*, written the *Methods* section that describes how you got the results, and explained the significance of the results in the *Discussion* (and an optional *Conclusions*), you are ready to write the beginning of the paper. The *Introduction* puts your work in a larger context. You may remember from chapter 7 that the *Introduction* is used to explain how a *local argument* of a paper fits into the *larger argument* in the scientific literature. You became familiar with some of the literature about the area you were researching as you were constructing your proposal, but you may find that you have to do a little more literature research before writing the *Introduction*. This is because, after you have finished off the *Discussion* and *Conclusions*, you may have a better idea of just what your local argument is, and it may now fit into different larger arguments within the scientific literature.

The *Introduction* is one of the most consistently structured parts of a scientific paper. No matter which branch of science you are reading in, you will always find a common progression from the general to the specific. The first sentence or two will orient the reader to the general field of research, usually emphasizing its importance. These sentences can sometimes seem obvious and boring to the person writing the paper, because they are often statements of what has been taken for granted by the researcher before they even started doing the research. But they are part of the argument that the paper is related to an important field of research that concerns a lot of people. The first sentence might give no new information to many of the readers who will be most interested in the paper, but it announces that the paper is relevant to their interests.

Subsequent sentences will typically discuss the theoretical background of the research, other research that has been done on the problem, and what this research has found. These sentences might be important to help someone new to the field understand what the key issues are in the field and what else has been going on in the field, and give them some references to other literature that has contributed to the larger arguments in the field. These sentences are also setting the stage for why the study being presented is needed or helpful.

The end of the *Introduction* will explicitly state what problem or knowledge gap remains in the field, and why the research to be reported in the rest of the paper helps solve that problem or fill in that gap in knowledge.

Let's consider how we might apply this overall plan to presenting a study on the effectiveness of different laundry detergents. (In the following fake *Introduction*, all of the cited references are fake, too.) We start with the most general statements, and some theoretical background:

> People have been washing clothes for thousands of years, using a variety of methods and chemical agents to help them remove soil and stains. In the last two centuries, homemade soaps and natural chemical aids have increasingly been replaced by industrial products.[1] These detergents are typically formulated to include a variety of different classes of chemicals including surfactants, builders, enzymes, chelating agents, and whiteners, and their exact compositions are often proprietary and not known to the consumer.[2] Surfactants are of primary interest to the removal of oil-based stains, as they form micelles that

can surround hydrophobic molecules and solubilize these compounds in water.[3] Detergents can contain a wide variety of ionic and non-ionic surfactants, and the formulation of detergents often involves trade-offs between the effectiveness of surfactants and their cost.[4]

Then we would have a summary of "previous research." Here we already start to hint at what sort of research might be needed to better address the general topic.

Given the variability in surfactant composition and the proprietary nature of detergent composition, the consumer must rely on empirical measures of detergent effectiveness. Television advertisements for these laundry products often show experiments in which one detergent is shown to be superior to another "leading brand" in cleansing a particular piece of clothing of a particular stain. For example, a recent advertisement shows Nuke-It removing grass stains from a t-shirt better than Ecorama or Hi-White, but the exact washing conditions, such as temperature, were not specified.[5] Other advertisements single out particular strengths of detergents, such as a recent radio advertisement by Ecorama, which claims "the best cold-water cleaning power of any leading detergent."[6] Some researchers have attempted to make more controlled studies of commercial detergents under more rigorous laboratory conditions,[7-14] but many of these studies do not use realistic stains, and many of the studies were done too long ago to incorporate brands currently on the market.

Note that, in this particular instance, some of the citations are not of the usual scientific literature; it would be great if we always had good scientific literature to cite, but sometimes this is not available!

The last part of the *Introduction* should then clearly state what gap in the research exists, and how the study being presented fills that gap.

Wash temperature can affect the effectiveness of laundry detergents, especially in the removal of oil-based stains. To date, no scientific study has been published that compares currently available detergents at a variety of temperatures with a typical oil-based stain. In this study, four leading detergents were used over a range of temperatures on the removal of lipstick stains from white t-shirts. Lipstick stains were chosen because they have a high pigment load, which allowed for easy quantification of results, and because the hydrophobic vehicle of the lipstick would present a fairly typical challenge for the surfactants contained in the detergents.

In these three paragraphs, we have gone from the very general—people washing clothes through the ages—to the very specific—lipstick stains on t-shirts. In getting from the very general to the very specific, we presented what is known, what has been done, and what problems haven't yet been solved, or what questions haven't yet been answered. Together, these present a good argument that our study on lipstick stains will contribute to the larger argument in the literature about the relative effectiveness of different laundry detergents.

13.11 ABSTRACT

The *Abstract* is a fairly new part of scientific papers; as we discussed in chapters 4 and 5, it wasn't until the mid-twentieth century that abstracts actually started appearing in journals next to the articles. Originally, the purpose of abstracts was to provide summaries of articles that scientists couldn't readily access. Now that so many papers are available electronically, their purpose has evolved; they tend to be written more as an aid to reading or even an advertisement for the paper.

Many books on scientific writing state that abstracts should summarize the entirety of a scientific paper, including the *Introduction*, *Methods*, *Results*, and *Conclusions*. But the emphasis given to various aspects of the paper will vary depending on the journal to which they are submitted. For years, my students and I have been examining abstracts in the *Journal of the American Chemical Society*, a traditional chemistry journal, and we find that most of the abstracts focus on just methods and results. They don't include much about the background or the interpretation of the results. It appears, however, that this is more of a convention in chemistry than it is in other disciplines.

Journals such as *Science* and *Nature*, whose readership spans a broad range of scientific fields, are at the other extreme. The abstracts in these journals often have several sentences of background and one or two sentences at the end that go beyond results and into the interpretation of results. This type of abstract helps readers from diverse fields to understand the context and importance of the research and is a good example of how the abstract has evolved to be more of an aid to reading or an advertisement for the article.

Even further in the direction of advertisement is the request of some journals for graphics or "visual abstracts" to put in the table of contents for the journal. With such a large number of scientific papers being published, it is increasingly important to grab the reader's attention with an easily accessible graphic.

Although the purpose of the *Abstract* has evolved, the *original* purpose of abstracts is good to keep in mind when writing an abstract. Imagine yourself writing an abstract for someone on the other side of the world who needs to decide if it is worth it to go through a lot of trouble and expense to get a copy of your paper. Putting yourself in that frame of mind will help you focus on what is most important to convey to the reader. Although it may reflect my chemistry bias, I think it's more important to get the most important *Methods* and *Results* into the *Abstract* than background or interpretation from the *Introduction* or *Discussion*. An abstract shouldn't be too long, and what you did and what results you got are still the most important parts of any scientific research.

13.12 TITLE

The two major issues you need to address in writing a good title are (1) how specific to make it and (2) the grammatical structure of the title.

There was a time when you could publish an article entitled "A Relation Concerning Barnacles" and get away with it.[5] That time was the seventeenth century. Since that time, imagine how many dozens, hundreds, or even thousands of articles have been published on barnacles! Now, any paper about barnacles is likely to be far more specific, such as "Distribution and orientation patterns of the pedunculate barnacle *Conchoderma* sp. on the swimming crab *Portunus trituberculatus* (Miers, 1876)."[6]

I suppose it is possible to make your title *too* specific, but I have rarely encountered this problem in the published literature or student work. The more information you can put in, the better; the only problem is when your title gets to be so long that it's too long to read aloud in a single breath.

The *grammatical* nature of the title is one of the areas of scientific writing that has, surprisingly, generated some controversy. In the late twentieth century, it became popular in biology journals, particularly in molecular and cellular biology, to use titles that were sentences, sometimes called "assertive sentence titles"[7] or "declarative titles."[8] An example of a randomly chosen declarative sentence title from *Cell* is "Mutant KRAS enhances tumor cell fitness by upregulating stress granules."[9] In this title, the subject is "Mutant KRAS" (a gene), the verb is "enhances," and the object is "tumor cell fitness." So, "Mutant KRAS enhances tumor cell fitness" is the skeleton of the sentence, and the rest of the sentence outlines what the authors believe is the mechanism by which this happens.

Does this title fairly summarize the conclusions of this article? The end of the discussion for this article is more nuanced than the title: "In principle, therefore, in a setting of mutant KRAS, SGs *may* have a distinct composition that, in turn, *may* impart a unique translational and post-translational stress signature. Clearly, *further investigation will be required* to understand whether such a signature could be predictive of overall tumor cell resilience and chemotherapeutic resistance." I have added italics to indicate those cautious words and phrases that we are used to seeing in discussions—words and phrases that emphasize the tentative nature of many new scientific discoveries.

It is much easier to find opinions criticizing the new style of declarative title than it is to find definitive explanations for its rise, or statements in favor of its use. Rosner thought the rise of such titles may have been the result of a desire to make science more of a product, with definitive conclusions for every publication.[10] Perhaps authors and editors thought a simple sentence might just be a clearer statement of the conclusions of the paper.

But, as Rosner[11] and others[12] have all pointed out, the conclusions of a single paper in the scientific literature are rarely definitive, and an overly confident title can promote a somewhat tentative and shaky conclusion into something that is repeated as a simple fact. Some have even suggested that the reduction of nuanced and uncertain conclusions into simple fact could lead to medical malpractice. Although I think this is unlikely, there are more fundamental reasons for avoiding the declarative sentence titles. The larger arguments in the scientific literature are logico-inductive arguments of considerable complexity—go back to chapter 1 if you've forgotten how many philosophers of science have struggled with the problem of how we know what we know! Usually the best scientists—the ones that you most want the good opinion of—recognize this and look favorably on titles that convey recognition that a single paper rarely "proves" anything.

Despite the editorials against declarative titles that I have cited here, a glance at the scientific literature shows that, although their use has declined somewhat, they are by no means gone, especially in the journal in which their rise was most prominent: *Cell*. In fields other than biology, these titles aren't so popular; in astrophysics, only 3 percent of articles surveyed in one study had such titles (called *verbal* titles in that research).[13] My personal advice would be to avoid the assertive sentence title or declarative title, unless you are submitting an article to a journal in which most authors use such titles.

If a title isn't a sentence, what is it? A more traditional title for an article discussed above might be something like "Stress granules in mutant KRAS cells as a possible cause of tumor cell fitness." This title is not a sentence, because there is no verb; there is only the noun "stress granules," and everything else just modifies that noun, describing why these granules are interesting. Many titles in the scientific literature are just nouns with a lot of modifiers.

13.13 PUTTING IT ALL TOGETHER

Scientific papers usually are written in pieces, and often are read that way as well, as we suggested in chapter 7. Nevertheless, it's a good idea to look over the paper as a whole when it is all put together in the proper order. Don't worry about "smooth transitions" between the sections—although these are often wanted in other kinds of writing, they are neither expected nor encouraged in the highly structured form of a scientific paper. It's more important that the right things are in the right places. If you find results in the *Methods*, or methods in the *Results*, or discussion in the *Introduction*, or discussion in the *Results*, get it out, and put it where it belongs. Scientific writing is less about "flow" and more about "sorting into bins."

Looking at the paper as a whole is also about seeing if it is complete, but without superfluous material. Are there results for which no methods were described? Is there discussion unsupported by any results? Does the information in the *Introduction* lead readers to think you are going to try to answer questions that you don't address with your research? Although the different material in the different sections has to stand alone to some extent, readers should be able to find connections among the materials in the different sections.

Finally, check all the details. Make sure your citations are correct. Make sure your figures and tables are properly formatted and labeled. Make sure all your figures and tables are referred to in the text.

Whether you submit your paper to a journal or to your professor in a class, expect that you will get back criticisms, either from peer reviewers and editors, or from your professor. You can read more about the peer review process in chapter 8. Regardless of where the criticism comes from—and regardless of how insulting, cruel, and infuriating it might be—relax. All scientists have suffered such corrections, insults, and hurtful words! Even if the criticisms are not presented in a constructive and positive manner, look for constructive and positive ways of interpreting them. Be courteous to those who criticize. Be ready to revise, and don't expect the revisions to be small or insignificant. Science is, after all, a social process, and responding to criticism and fixing error is pretty much the way science has progressed over the last five centuries.

FOR FURTHER STUDY AND DISCUSSION

1. Find an article in the literature of your discipline that has a combined *Results and Discussion* section. See if you can determine which sentences, or even parts of sentences, would belong in a *Results* section, and which would belong in a *Discussion*. Are there good reasons for merging the *Results* and *Discussion* in this paper? What are they?
2. Look through a popular journal in your discipline. Are all the sections about how the experiments were done called the same thing? Are there common grammatical or stylistic conventions in these sections? How many references are there in these sections?
3. Find an article in the literature that has a *Discussion* section separate from the *Results*. Classify each paragraph as either (a) commentary on the suitability of the method and quality of the results, (b) comparison of the results with those published elsewhere, (c) practical implications of the results, (d) theoretical implications of the results, (e) future research suggested by the results and their interpretation, or (f) other.
4. Look at two or three *Introductions* in the scientific literature. Do they follow the usual pattern of starting with very general statements about a problem and ending with identification of a particular knowledge gap that the research seeks to fill?
5. Look at some abstracts in a popular journal in your discipline. To what extent do the abstracts contain introductory material, methods, results, and interpretation? What is the shortest abstract you can find? What is the longest you can find? Is the shortest too short, or the longest too long?

ADDITIONAL READING

Day, R. A. *How to Write and Publish a Scientific Paper* (Oryx Press, 1994).
    An older work, now in its eighth edition. Its historical perspective was an inspiration for the historical emphases in this book.
Gilpin, A. A. & Patchet-Golubev, P. *A Guide to Writing in the Sciences* (University of Toronto Press, 2000).
    A nice, short work that covers the basics.

Hofmann, A. H. *Scientific Writing and Communication. Papers, Proposals, and Presentations* (Oxford University Press, 2016).

Very detailed and comprehensive, including writing tips for those for whom English is not their first language.

Penrose, A. M. & Katz, S. B. *Writing in the Sciences. Exploring Conventions of Scientific Discourse* (Pearson Longman, 2010).

I like this book for the way it is more descriptive than prescriptive and takes a rhetorical view of scientific writing. This book definitely helped shape my course in the early years.

Robinson, M., Stoller, F., Costanza-Robinson, M. & Jones, J.K. *Write Like a Chemist. A Guide and Resource* (Oxford University Press, 2008).

Covers problems that are endemic to writing in chemistry. A good reference book for chemists to have around.

Tufte, E. R. *The Visual Display of Quantitative Information* (Graphics Press, 2001).

A classic text about making effective graphs. Many of the examples are above and beyond what you will see in scientific journals, but the work is nevertheless inspirational.

CHAPTER FOURTEEN

ORAL AND POSTER PRESENTATIONS

LEARNING OBJECTIVES

After reading this chapter, you will be able to:

- explain the role of oral and poster presentations in the history of scientific communication
- identify when, and why, poster sessions became a popular forum for scientific communication
- explain the differences between the structure and style of written scientific communications, oral presentations, and posters
- explain how graphics for oral presentations and posters should be different from those for published scientific papers

14.1 HISTORICAL PERSPECTIVE

Oral presentations of research have been a part of science for at least as long as written and published reports. Modern oral presentations of research, however, are in some ways quite different from what used to occur in the scientific societies and academies of the past.

In the seventeenth century, the *Philosophical Transactions* of the Royal Society was filled with papers and letters that were read aloud at the meetings of the Society (see chapter 4). Sometimes these were presented by the authors themselves; at other times, scientists unable to come to the actual meetings sent letters to be read by the secretary. Likewise, the *Mémoires* of the Paris Academy of Sciences (see chapter 8) consisted of papers that were presented at meetings before being published. In the event that there was a letter read from a distant correspondent, the oral presentation and the text were probably the same. For papers delivered by the authors, we don't really know to what extent the speakers read fully prepared texts or spoke extemporaneously. For the *Comptes rendus*, the principal publication of the Paris Academy after 1830, there must have been rather finished texts ready to be published, because they were due to the secretary by the end of the meeting in which they were read, for publication before the end of the week. Regardless of how much improvisation and modification might have occurred, however, through much of the history of the scientific literature there was a direct one-to-one correspondence between presentations and publications.

Modern presentations are different in that they are frequently *not* the same as written letters and articles about the same research. There are, of course, books called conference proceedings (which we looked at in section 3.4), where the published text is quite close to what was presented. However, there are many conferences now where no text is published—the primary purposes of the conferences are to allow scientists to share work in progress, get feedback from other scientists, and, essentially, advertise the science they are doing. Scientists also travel to give seminars at companies and universities, and the presentations they give at these locations may be retrospectives of a lot of older published papers as well as new work that is still in progress. Students finishing graduate, and even undergraduate, degrees are often asked to give oral presentations of their work as a final exercise for their degrees. Scientists also make presentations to companies and universities that might hire them; these are usually referred to as "job talks." In fact, many of the oral presentations that modern scientists do are referred to as "talks," and this conveys the sense that they are typically less formal than papers, less rigorously structured, and less likely to correspond to particular published work.

14.2 THE STRUCTURE OF ORAL PRESENTATIONS OF RESEARCH

In structuring a talk, the modern scientist often won't follow the *Introduction, Methods, Results,* and *Discussion* format of the modern scientific paper that we introduced in chapter 4, and that we used to structure our work in chapter 13. Remember, that format itself is a relatively new invention, which only began to take over the scientific literature in the mid-twentieth century. One of the main advantages of the IMRAD format is to make it relatively easy for scientists to get what they want out of a paper without reading the whole thing. With a talk, however, it is important to keep as much of your audience awake and engaged as possible, for as long as possible. Once someone in the audience gets lost or confused, you might well have lost them for the rest of the talk.

Being flexible in structure is especially important when talks are long and involve many different experiments. Imagine if you had a fifty-minute talk summarizing all the work done in an academic laboratory of half a dozen graduate students over the course of four years. There would be lots of experiments, lots of different methods, and lots of different results. If you organized this talk in a strict IMRAD format, you might well have ten minutes of introduction, followed by a ten-minute presentation of five different methods, followed by fifteen minutes of the results of different experiments done with these methods. Ten minutes into the discussion of these results, the audience might be having a hard time remembering the methods or results you are discussing. It is far better to do what humans have been doing around campfires since *before* there was writing and reading: tell stories.

In previous chapters, I wrote that papers don't describe the actual chronological development of the research—but that may be something worth considering in giving a talk. Maybe you thought that this one method would give you this one result, but it's not what you got. So, you tried a new hypothesis, a new experiment, and got a different result, and then you thought about the problem differently. Then a new experiment was tried; the results may have confirmed your new hypothesis or forced you to think in still different ways. If you tell this *story* about your research, you will be going around in a spiral: methods, results, discussion, new methods, new results, more discussion, and so on. Your audience will be able to see each episode on its own, and you will have conveyed some of the surprise and excitement of the research.

Be careful, though, not to get too caught up in this spiral. Not every little modification of every experiment deserves its own episode! You are still going to have to edit reality down a lot; if you were to present all the details of research done by six graduate students over four years, you would have twenty-four person-years of trials, failures, and successes to report. The stories you tell in a talk will be, in some ways, a little bit fictional because of this editing; this is okay as long as you are bend-over-backward honest and non-fictional about the key experiments, results, and conclusions. Your audience will understand that a lot more stumbling about may have occurred that you did not discuss in detail.

It is important, too, to frame the research properly. You cannot abandon the *function* of the *Introduction* of a paper, even if you don't have a section of the talk labeled as the *Introduction*. You have to orient your audience to what was already known about the research questions before you started. Likewise, you will need to spell out the conclusions of your research, even if there is no formal section of your talk labeled *Conclusions*.

In summary, while we may not want to strictly follow the IMRAD format that we use for papers, we do need to organize the talk well and give introductory information and concluding statements. If the talk covers several different areas of research, it will also be necessary to clearly divide up the talk so that the audience doesn't think your results from one sort of research pertain to an entirely different field of research that you finished talking about five minutes ago. Remember that many people in the audience will have little or none of the background information you have, and won't have thought about your research problem nearly as much as you have. Organize the whole thing into small, easily digestible bites. Make sure you clearly signal to the audience, both in what you say and in the visual aids you use, where one bite starts and another begins. Repetition of important points may sound boring to you, but people's attention can wander, and you want to make sure as much of the audience as possible is following along.

14.3 VISUAL AIDS

In early meetings of scientific societies, it was not unusual to actually demonstrate simple experiments at scientific meetings. Preserved specimens or large illustrations might also be shown. The modern scientist, however, has a much more powerful tool to use for displaying images and data: projection of computer-stored images. As we have seen throughout this book, tables, graphs, and other images can convey a lot of information very efficiently and accessibly.

The increasing use of displayed images has, of course, a history. The first use of a wall-mounted blackboard to convey information to a class of students is generally dated to 1801, and blackboards were probably adopted for scientific talks soon afterward. As late as the 1980s, some PhD students in organic chemistry were still required to give "chalk talks" using blackboards as a part of their work to earn their degrees.

In the period following World War II, both thirty-five millimeter (35 mm) slide projectors and overhead projectors came into wide use for presentations. When I was in graduate school, in the 1980s, fields where a lot of visual information needed to be shown usually used projected slides made from 35 mm photographic film; fields where mathematical expressions were more central tended to use overhead projectors. I was a projectionist for seminars where I did my PhD; organic chemists and inorganic chemists giving seminars usually brought along a stack of slides; physical chemists brought along overhead transparencies on which they had written all the equations that represented their research. Slides required some development time, but could include detailed visual data such as photomicrographs, spectra, diagrams of apparatus, maps, and well-drawn chemical structures. Overhead projectors, on the other hand, utilized write-on transparencies that could easily be corrected or revised if mistakes were found in equations or if slight modifications were made to the theory expressed by the equations.

Both of these technologies are obsolete now; video projection of computer-generated images has pretty much taken over. Many popular applications for generating such images, such as Microsoft PowerPoint, Google Slides, Apple Keynote, and others, still refer to the individual units of a presentation as "slides," evidence of the old film-based technology. But since the 35 mm slide only dominated scientific presentation for fifty years or so, we can hardly expect its influence to last far into the future as people imagine new ways of conveying visual information. However, at the time I'm writing this, "slides" are still what people call the individual layouts of words and data that are projected before an audience, so that's what I'll use in this chapter.

One thing that you should avoid with any of these applications is to let the premade templates in the applications dictate your content. There may be different types of slides in the template premade with lists of bullet points and places for charts, but these should be abandoned in favor of blank slides. Let your own ideas of what you want to communicate dictate the arrangement of your slide.

14.4 HOW MUCH TEXT?

The short answer to that question is "Not too much." It is certainly possible to put your entire talk, every word of it, on successive slides, and just have the audience read it. At the other extreme, one could only put graphs and pictures on slides, and convey all other information through speaking. In practice, most scientists end up somewhere between these extremes. We now know that the most efficient way to communicate ideas to people depends upon the person; some people learn better by reading rather than listening, and others learn better by listening. Having *some* words on your slides helps those people who prefer visual input to auditory input. On the other hand, you don't want people to be so busy reading that they don't have time to listen to what you're saying.

Another reason for having words on your slides is that they can serve as your notes when you give the presentation. I've rarely seen scientists giving talks using notecards; if they need to be reminded of what comes next in their talk, they advance to the next slide and the words or pictures on that slide will remind them. But you don't want your slides to be an outline of your talk—you want to focus on presenting images that convey your methods, results, and conclusions. A good guide to how much text should go on slides is that you should have good reasons for every word. Below are some of the most frequent situations where text is often a good idea, and the reasons for the text.

A title slide. You may have someone introducing you at the beginning of your talk, or you may be introducing yourself. People who may want to contact you in the future about your work may want to write down your name and what institution or business you come from. You can't be sure that people will know the correct spelling of your name from how it's pronounced. A title slide with the title of your talk, your name, your institutional affiliation, and maybe some contact information should be displayed for long enough for people to get down whatever information they need.

Outline slides. These are especially useful if your talk is long and has lots of different parts. Giving your audience some idea of what different things you're going to be talking about helps them stay oriented as you go through your talk. I would advise against using an outline slide if the outline just consists of "Introduction, Methods, Results, and Discussion," however. As we discussed in the last section, you'd only strictly adhere to this structure if you had a fairly simple talk, and in that case an outline probably isn't necessary. Scientists will naturally recognize the usual standard paper format, and an outline slide reiterating the IMRAD convention isn't going to convey any information to them.

On the other hand, if your talk is long (more than 20 minutes), you not only should have an outline slide but should display a duplicate of that slide every time you transition from one major topic to another, just to remind people how what you are saying fits into the overall organization of the talk. Use of color or bold text to highlight where you are in the overall outline makes it that much easier for people to follow along.

Lists of major questions or conclusions. Often, in scientific research, it isn't just a matter of asking a simple question and getting a simple answer. There are multiple outstanding questions, and a set of experiments can often constrain the answers in several different ways, even if final answers aren't arrived at. A bulleted list of questions or conclusions can emphasize to the audience that there are multiple questions and multiple conclusions. In listening to a talk, many people may focus on just one main question or one main answer. That's fine; we all have limited ability to attend to and retain information. But the visual of a bulleted list will help your audience remember that there was more to your talk than the one question or answer they most easily remember.

But be careful—if your lists have more than four or five items, they're too long. If you really have that many questions, or answers, your talk probably needs to be reorganized so that you're never dealing with a few answers or questions at a time.

Technical terms. If you don't think everyone in your audience knows the terms you are using, you should include them on your slides as well. You don't have to define them on the slides—you should be *talking* about what these terms refer to, not having people read all that information—but if the terms or names are something that people might want to write down or look up later, they ought to be up on the slide so that people at least know how they're spelled. The terms can be placed in diagrams or figures showing experimental apparatus or data, or in explanatory figures showing the main objects of study such as a particular molecule or organism.

Titles for graphs and other figures. These aren't going to be the same as the figure captions that you would write for a published paper. For a paper, you have to give enough information in the caption so that the graph can stand on its own; readers are likely to want to know enough to understand the data and how it was obtained without referring to the text of the paper. In a talk, however, you will be there explaining the data on the graph, and the audience should be listening to you rather than reading a lot of details about the data. The title is there just so that, if someone gets lost in your explanation and wonders, "Now, what is this data?" they can at least be reminded by the title what the data is, and what it is supposed to mean. There's no reason to number the figures, however; a figure in a paper needs to be labeled as "figure 1" so that references from the text can be correlated with the right figure. In a talk, though, you're right there, explaining the figure.

Titles for tables. The situation for tables is similar to that for figures—not nearly as much information is necessary as in a paper, since you'll be explaining the table yourself while your audience is looking at it. Because tables present particular challenges in talks, however, we will be explaining just what to do with tables in the next section.

References. If, in presenting the context of your research, you show data or other figures from published work, you should definitely reference it. Failing to do so is plagiarism! And don't make the reference text too small, either; some in the audience will want to look up these articles or books. Unfortunately, people in the audience frequently don't have enough time to copy the whole reference into their notes as the speaker moves ahead to the next slide, but at least some key information can be taken down. If you're in the audience, just try to take down the first author, a year, and a journal—with that information, *Google Scholar*, *Web of Science*, and other databases we discussed in previous chapters can probably get you to the reference.

The assertion-evidence approach. There has been a recent campaign to have scientists and engineers adopt an *assertion-evidence approach* to developing talks and slides for those talks. Much of what its proponents advocate is similar to what I have always recommended: don't just outline your talk in bullet points on your slides; don't use words that are not necessary; and focus on visual presentations of methods, results, and models. The one thing that is different is that this approach demands that you have an assertion—a complete sentence—that applies to the graphic which takes up most of the slide. These assertions are in some ways similar to the "assertive sentence titles" that we discussed in the last chapter. The cautions we discussed there may apply to some of the assertions on slides, too, but in the case of a single slide, assertions may be no more than the sort of sentences you might put in a methods or results section of a written paper, and in many cases the "evidence" is no more than an illustration of a method or model. A study comparing audience comprehension suggests that use of the assertion-evidence approach in a presentation improves the presentation's effectiveness.[1] The control in this study, however, was a talk which used "slides ... modeled after the structures most easily created using Power Point default settings...."[2] These settings favor the use of text to communicate ideas—which was never something I was encouraged to do back in the days before PowerPoint! So, while the assertion-evidence approach is described as innovative and dramatically different, I would say it really harks back to the advice given to graduate students in the pre-PowerPoint era to minimize the amount of text on slides. There are definitely some good features of the assertion-evidence approach, though: it forces you to be sure that there is a central idea to each slide, and the assertion helps you, as speaker, remember what the slide is really about.

14.5 TABLES AND FIGURES FOR PRESENTATIONS

You shouldn't just take the figures and tables you made for your paper and insert them into the visual aids for your presentation without thinking critically about the differences between papers and talks. Remember, what often drives decisions in the world of publishing is minimizing the space an article takes up. At a talk, however, a really complex figure with a lot of information in it is going to overwhelm your audience. Some people will be trying desperately to make sense of the figure and stop listening to what you are saying; others may just give up on getting anything out of the graphics at all.

For talks, tables and graphs must be disaggregated and simplified. Let's look at table 14.1, taken from an article I wrote for my PhD thesis work. This was a great condensation of a lot of information for publication, but it would make a terrible slide for a talk.

Table 14.1. Binding Constants, K_{CO_2}, of CO_2 to Co and Ni Tetrazamacrocycles in $(CH_3)_2SO/0.1$ M $TBA(ClO_4)$. Effects of redox potential, steric configuration, solvent, and alkali metal cations on the binding of carbon dioxide to cobalt(I) and nickel(I) macrocycles. Reprinted with permission from Schmidt, M. H., Miskelly, G. M. & Lewis, N. S. *J. Am. Chem. Soc.* **112**, 3420–3426. Copyright 1990 American Chemical Society.

Complex	Ligand	$E°'$ $(M^{II/I})^a$	K_{CO_2}, M^{-1}
$[Co(Me_4[14]1,3,8,10\text{-tetraene})]^+$	1	−0.76	$<4^b$
$[Co(Me_2[14]1,3\text{-diene})]^+$	2	−1.31	$<4^b$
$[Co(Me_6[14]1,4,8,11\text{-tetraene})]^+$	3	−1.42	<4
$[Ni(Me_6[14]ane)]^+$	8	−1.69	<4
$[Co(Me_8[14]4,11\text{-diene})]^+$	4	−1.70	7 ± 5
meso-$[Co(Me_6[14]4,11\text{-diene})]^+$	5	−1.74	$(2.6 \pm 0.5) \times 10^2$
d,l-$[Co(Me_6[14]4,11\text{-diene})]^+$	5	−1.74	$(3.0 \pm 0.7) \times 10^4$
$[Co(Me_4[14]1,8\text{-diene})]^+$	6	−1.80	$(1.0 \pm 0.3) \times 10^5$
$[Ni([14]ane)]^+$	9	−1.89	irreversible
$[Co(Me_2[14]1\text{-ene})]^+$	7	−2.00	irreversible

[a] Measured vs ferricinium/ferrocene internal reference.
[b] An irreversible reaction with CO_2 occurs upon reduction to Co^0.

When I presented this same information in talks, the information in this table was spread over *many* slides. There were several reasons for doing this. One goes back to the organization of the talk; the talk was more of a story I told about my research, and I wasn't trying to show all the results in a clearly delineated *Results* section. In my story I first discussed what I learned about the effect of oxidation-reduction potential ($E°'$) on the binding of carbon dioxide to these complexes. Therefore, I first only showed those lines of the table relevant to that aspect. Then, in telling the story of the effects of stereochemistry on the binding of carbon dioxide, I had a separate slide containing only the data for the *meso*- and *d,l*-isomers of the cobalt complex of ligand 5. Another slide showed the data that compared the nickel complex with some of the cobalt complexes.

The formatting of each of these tables was different, too. The bold numbers in the "ligand" column of the published table referred to chemical structures in a separate figure in the paper. That was necessary because not all chemists would recognize the names of the complexes in the first column of the table. But in giving a talk, I didn't want to go back and forth from the tables to a slide with the structures. Instead, I just put pictures of the complexes directly in the tables I made for the talk. Figure 14.1 shows a table, formatted as a slide, that is more appropriate to a presentation. This slide just focuses on the results for the two different stereoisomers of one complex. The chemical structures in this table not only save the audience the work of decoding the chemical formulas but also give the speaker the opportunity to show just what the stereochemical differences are between the different isomers.

Notice that I have made other changes in this data table. The reference potential, which was carefully specified in footnote (a) to the table, is now abbreviated in the header. The error limits in the binding constant values are not included, and these binding constants are no longer in scientific notation, making the differences more visually evident. It is true that some of these changes make the table less complete, but in a presentation, you can't expect your audience to absorb every detail of your research. You want to keep your audience with you, and you want them to walk away with the main ideas and conclusions from your research. This slide tells them, with very few words, that changing the stereochemistry changes the binding constant by two orders of magnitude, even as the oxidation-reduction potential remains the same. (If I had taken the assertion-evidence approach to this slide, I would have worded the text a little differently: "Stereochemistry affects the binding of carbon dioxide to metal complexes.") The audience can see what these complexes look like, and how the stereochemistry is different. If they are really interested in details such as the possible errors in the values obtained for the binding constants, they can ask questions after the talk, contact you by email, or read the publication.

The situation with figures is similar. Sometimes we put together graphs with many sets of data from many experiments. If you can gradually build up the graph one data set at a time, or show a number of graphs, each of which compares only a few data sets, you will have a better chance of keeping your audience with you and not confusing them.

Effects of Stereochemistry on Binding of Carbon Dioxide to Metal Complexes		
Complex	$E^{\circ\prime}$ vs. Fc^+/Fc	K_{CO_2}
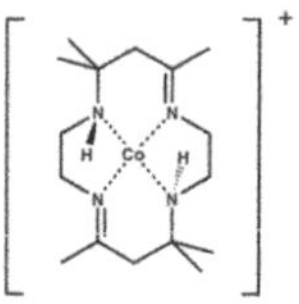	−1.74 V	260
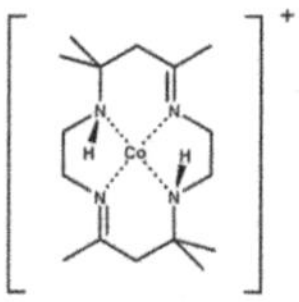	−1.74 V	30,000

Figure 14.1. A slide consisting of a table illustrating just a portion of the data in table 14.1.

14.6 SLIDE STYLE

This is an issue that I'm reluctant to write about, as it involves matters of taste, and taste evolves with time. I came of age when people were formatting simple slides on white paper using the limited computer capabilities of the time, and taking pictures of these black-and-white graphics with black-and-white line-copy film to save money. The results ended up looking a lot like figure 14.1, and that simple style is what I am still most comfortable with.

After PowerPoint and other programs became available, both the developers and the users became fascinated with the possibilities of greater computing power. Changing the color and design of backgrounds and text became easy, and animated text, sound effects, and music became possible. When I first started having students do presentations, students avidly adopted these more advanced features, but their presentations actually suffered as a result. The pace of the talk was dictated by how long the sound effect or animation lasted, not by what was most appropriate for what was being presented. The end result was often that it would seem that the computer, not the student, was in charge of the talk.

Fortunately, a lot of that initial enthusiasm has worn off. The use of "themes"—styles that have shaded or textured backgrounds and special typefaces—is still pretty common, but the use of animation is sparing, and the use of sound is almost nonexistent. Animation is fine if it really shows something of scientific interest. An animated graphic showing the effects of plate subduction at continental margins or the signaling pathway in a cell is perfectly justified, but I would avoid animating text. It looks like a gimmick. As I wrote in chapter 11, scientific communication isn't about expressing yourself or showing how clever or creative you are in the presentation of your data and ideas. If your science is sufficiently new and interesting, your audience will devote all of its attention to understanding the science you are presenting rather than appreciating fancy designs and special effects.

Although many scientists now routinely use colored, designed backgrounds, most of my presentations still have white backgrounds and mostly black, or very dark-colored, text. This may be excessively conservative, but I like to think that the high contrast makes the slides easy to read and understand.

14.7 TALKING THE TALK

After spending a lot of time putting together the slides for your talk or presentation, it is easy to think you know the talk well enough to give it. I've made that mistake before. There's a big difference between silently thinking through your talk and actually having the right words come out of your mouth at the right time. Practicing your talk in front of an audience is the best way to make sure you can talk your way through it. Practicing it without an audience is helpful, too, but people talk differently when there is an audience listening to them. It's okay if your practice audience is small and quite different from your eventual intended audience; it's the presence of people that is important. Of course, if the people know enough about your science to ask intelligent questions or give relevant comments, that's even better.

The actual speaking style for scientific talks these days is not very formal. You don't want to be peppering your talk with swear words and things only people under the age of thirty would understand, but you do want to come across as an open and friendly colleague who's interested in sharing results. Making your presentation into a formal speech may remind people too much of politicians, and politicians don't generally cause an audience to be very trusting. Remember that *honesty* is one of the highest moral values that scientists hold, and you should speak in a frank and honest manner. Finally, don't talk too fast. That can be hard, because often you will have a limited time for your talk and a lot of science to present. Nevertheless, it's better to make sure your audience leaves your talk with a good understanding of one good piece of your science rather than an incomplete grasp of bits and pieces of everything that you've done.

14.8 POSTER PRESENTATIONS

The poster presentation is a relatively new development. The rapid expansion of publicly funded science after World War II meant that, by the 1960s and 1970s, there were too many people wanting to give oral presentations of their research at conferences. Instead of restricting who could present, large scientific organizations gave scientists the option to summarize their presentations as posters. In this way, many more scientists could get their work seen by interested audiences.

Poster sessions have become increasingly popular, especially as a way for undergraduate students involved in research to practice their presentation skills. A major national meeting of a scientific society may have huge ballrooms in convention centers filled with hundreds of people presenting posters. Typically, there are aisles of corkboard room dividers upon which the posters are hung with thumbtacks. The presenters often—but not always—stand next to their posters so they can answer questions or give explanations or more details about their research to the audience that wanders by.

Poster sessions are of relatively recent origin, so there is still a lot of confusion about how best to present a poster. There are a lot of people with opinions, but beyond some basics, there isn't always agreement. How you present your research on a poster will depend on the amount of time your poster will be visible, whether you will be present during all of that time, how general or specific the poster session is, and how many other poster presenters you will be competing with for the attention of the audience.

If you will be present next to your poster for the entire time of the poster session, you can think of the poster *more* as an oral presentation. If the poster will be on display for hours, or days, and you won't always be standing next to it, then you should think of it *more* as a paper or article. But in either case it won't be either as complete as a published paper or as easy to digest as the story you would tell in a talk. It's going to be some sort of compromise, and it may be something of a frustrating compromise if you want it to substitute for either an article or a talk.

Perhaps a less frustrating, and more productive, way of thinking of the poster is as an *advertisement*. Think of people walking away from your poster anxious to finally see the research as a published paper or talk in the future. Tantalize your audience with some graphs or other figures of the data you've collected, and a statement about what you think they mean. Understand, however, that you may not be able to present a complete argument for your interpretation without talking with your audience and answering their questions.

The idea of a poster as an advertisement gives us some idea of how we should put the poster together. The audience at poster sessions have dozens if not hundreds of posters they could look at, and they won't look at all of them. You want them to look at yours. The one thing that is sure to scare away your potential audience is too much text. Amid all the excitement and bustle, the last thing people want to do is to stand in front of your poster for ten minutes reading a lot of careful scientific writing. As I recommended for the slides you prepare for talks, you need good reasons for every word and sentence of text that show up on your poster. You're going to have more text on a poster than on slides for presentations, but always consider whether what you are tempted to write could be better conveyed with graphics, or could be left for you to explain in person. If you are working on a paper at the same time that you are working on your poster, it may be very tempting just to copy a lot of that text right into your poster. Resist this temptation; critically examine every sentence and ask yourself, "Do I really need this on my poster?"

The idea of a poster as an advertisement also suggests that the organization of the poster doesn't have to follow the IMRAD format. Many resources on how to put together a poster do suggest following this format, and it is worth considering why. One reason to do so is that, as with reading a paper, people may want to skip around, starting by looking at your conclusions, or focusing on your methods. Another is that participants in poster sessions are often students and other junior scientists, and senior scientists tend to want to see their juniors "following the rules," even if the rules don't always make perfect sense in every context. So, while you should consider different organizational structures for your poster, it may be safest, and easiest, to stick with a traditional structure.

Within this overall framework, however, there are some options. Perhaps you can replace the *Discussion* with a *Conclusions* section; a real in-depth *Discussion* will probably require too much text. The *Methods* section doesn't have to cover every detail; you need only communicate your general experimental strategy to your audience. Depending on the methods, you might be able to communicate a lot with some simple diagrams. You should consider substituting paragraphs with simple bullet points where possible.

The assertion-evidence approach, discussed above for talks, might also be applied to posters, especially those that are closer to talks than papers. Instead of a separate *Results and Discussion* or *Conclusion*, you might just want to post an assertion above a graph of data collected. Posting "Stereochemistry affects the binding of carbon dioxide to metal complexes" above a table like that shown in figure 14.1 may be all you need to do to show and interpret that set of results.

Most people agree that you don't need an abstract on a poster. Think back to what an abstract is: it's a brief summary of your research, focusing mostly on methods and results, that helps people figure out whether it is worth retrieving and reading your article. It's an advertisement. What is a poster? It's a somewhat bigger advertisement, with more graphics. An abstract on a poster, then, is somewhat redundant. Often, too, a conference program will already contain an abstract of the poster, one that was submitted when the presenter applied for inclusion in the poster session. Viewers can read the abstract—or may already have read the abstract—in the program.

Another thing that is different about the text for a poster is that you don't want *nearly* as many references as for a paper. I would say a maximum of three or four. People wandering by your poster and taking a look at it are unlikely to write down or look up many references. If your introduction summarizes what has been done in the field previously, you'll need some references. If your introduction seems to require more than three or four references, though, your introduction is probably too long or complex. Yes, there may be some very interested viewers of your poster who want to find out more about the context of your research—but then you might want to come prepared with more references on paper that you can hand to the people who want to know more. Once again, think of the poster as an advertisement.

14.9 POSTER GRAPHICS

Just as a poster is somewhere between a talk and a paper, so, too, your graphs and figures should be somewhere in between those for a talk or paper. As in a paper, space will be at a premium, so combining as much data as possible onto one graph or table will be to your advantage. But, as with a presentation, the attention span of your audience will be limited, so you need to make sure your tables and graphs are as clear and as simple as possible. Obviously, these suggestions are opposed to each other, so you are going to have to work out a compromise. Think carefully about how color might be used to simplify the interpretations of your figures. Think critically about whether all the columns of data on your table really are important enough to put on the poster; if you can lead the viewer to your overall conclusion without them, leave them out!

Graphics on a poster don't just convey information, though; they also draw the attention of the people wandering by. Think about this in choosing what graphics to include. While graphs of data convey a lot of information, more explanatory figures like diagrams of apparatus, chemical structures, or pictures of organisms can bring in viewers who recognize such images as things that they're interested in.

14.10 POSTER LAYOUT AND DISPLAY

Once you have determined what is going to be on your poster in terms of both text and graphics, you must arrange these elements on the poster. Posters can be formatted with a variety of computer applications including Microsoft PowerPoint. You will have to know in advance what the acceptable dimensions for your poster are—these will usually be specified by the organizers of the conference or meeting at which you are presenting your poster. Posters that are wider than they are tall ("landscape" view) are often best organized into three or four columns; posters taller than they are wide ("portrait" view) are best with only two or three columns, or organized in rows. Some people try to have the flow of their poster wander back and forth across the poster, but most people will look at a poster the same way they read a book: top to bottom, left to right. Think of the columns you break the poster up into as pages. As with a lot of other scientific communication, you don't get extra points for cleverness; your goal is to make people think as much about your science as possible, and as little about your graphic design skills as possible!

The top margin of a poster is usually used for the title, which should be big enough for people to read from a distance. Authors are included as well, often with an asterisk for the author who is actually doing the presentation. This is different from published papers, in which the asterisk usually indicates the "corresponding author," the senior researcher who is in charge. Very often the insignias of the institution and the agencies funding the work flank the title in the top margin.

The process of assembling or printing posters changes constantly with new technologies. In the early days of posters, individual sheets of paper with text, figures, and tables were often pasted on a poster board or tacked to the corkboard provided by the conference organizers. Large-format color printing came along in the 1990s, enabling precise layout of the whole poster and easy setup. Large posters, however, need to be transported to the conference site rolled up in large cardboard tubes, and some people find this to be an inconvenience and one more piece of luggage that needs to be worried about when going through airports. Recently, a number of businesses have opened that will print your poster on cloth—cloth that can be folded and put into a suitcase. Some poster sessions now also allow "electronic posters," posters that are directly displayed on a large screen from a computer.

14.11 SUPPORTING YOUR POSTER

As I have mentioned, often you will be standing next to your poster during a poster session. You will be filling in all the details and explanations that don't fit into the limited space for text and graphics that you have been given. How to do this most effectively will depend on your audience, and it can sometimes be hard to judge your audience. At one moment, you might have someone who knows very little about your field of research, and they would like nothing more than for you to guide them through the poster step by step, essentially giving a mini-talk, with the poster serving as a visual aid for your talk. At the next moment, you might have someone who knows your field of research better than you do and can, at a glance, zero in on the part of the poster they are most interested in. This latter sort of person might regard a mini-talk as irritating, as they don't need all the explanation and want to quickly find out what they want to know and move on to the next poster.

The best way to deal with this is to politely ask, "Would you like me to guide you through this poster or would you rather just view it on your own and ask questions?" If they say, "I'll just look," then say that you'd be happy to answer any questions and let them stare. You may feel awkward asking this question but probably not as awkward as they would feel if they felt they had to listen to your mini-talk when what they really wanted to do was to move on to the next poster. If they want you to lead them through the poster, give them a mini-talk. Of course, this means you should be prepared to give a mini-talk, up to and including practicing it on friends or coworkers.

To add to the complications, in a crowded poster session you will also have people walk up to your poster when you are in the middle of a mini-talk to people who got there minutes earlier. Do you stop and bring them up to speed with where you are in your mini-talk? I wouldn't. They might be the sorts of people who don't want the whole mini-talk anyway. Finish your talk as you normally would to the first audience and then ask the latecomers if they have any questions or would like you to start over. It may not sound particularly graceful or efficient, but poster sessions themselves aren't particularly graceful or efficient. The best way to handle poster session audiences is to be flexible and give them the opportunity to tell you how they would prefer to learn about your research.

One thing that has become popular at some conferences is to hand out 8.5 × 11-inch reproductions of the poster, created from the same file as the poster itself. This allows the audience to walk away with a reminder of the research you were advertising without actually having to take notes. It's also a good way to make sure you don't have too much small text; if a poster is shrunk from 34 × 44 inches to 8.5 × 11 inches, 24-point type will shrink to 6 point! An alternative is to have flyers that summarize the main points, with some of the key graphics, arranged in a way that is better suited to the 8.5 × 11-inch format. The latest approach to providing more information is to have a QR code on your poster so that viewers can use their cell phones to access Web pages about your research.

Regardless of how your audience wants to interact with you or your poster, you should of course be friendly, helpful, and polite. Let your enthusiasm for your research shine through. If your audience seems receptive to more interaction, try to learn something about them and their research as well. They are more likely to remember you and your research if you show an interest in theirs.

FOR FURTHER STUDY AND DISCUSSION

1. Attend a seminar or talk by a visiting scientist. Answer the following questions about their presentation: Was the talk organized around traditional divisions of *Introduction, Methods, Results,* and *Discussion,* or did they weave different elements throughout the talk to create a story? Which slides particularly grabbed your attention? Which were too full of information?
2. Find some posters on your campus that present research. Choose two posters and answer the following questions about each: Did the poster use traditional divisions such as *Abstract, Introduction, Methods, Results,* and *Discussion?* Were these helpful? What graphics were used? Were there too many or too few graphics, in your opinion? What drew you to the poster? What put you off about the poster?

ADDITIONAL READING

Hofmann, A. H. *Scientific Writing and Communication. Papers, Proposals, and Presentations* (Oxford University Press, 2016).
Hofmann's book includes chapters on posters and oral presentations.
Penrose, A. M. & Katz, S. B. *Writing in the Sciences. Exploring Conventions of Scientific Discourse* (Pearson Longman, 2010).
The chapter in this book on "Preparing Conference Presentations" covers both oral presentations and posters.

CHAPTER FIFTEEN

CLOSING THOUGHTS

In this book, we started with big questions about what humankind is, what science is, and how science has succeeded in finding out more about the physical world. We looked at how scientists can conduct themselves in ethical ways, both as humans in general and more specifically as members of the scientific community. We discussed where ideas come from and how to shape these ideas into experiments and research that other people will care to read about. We learned how to find and read about the research of others, and how to most effectively share our own research with others. It may seem a bit disappointing, then, that I ended the last chapter with a section on poster presentations, which I said you might want to think of as advertisements of your research. Because of space limitations, posters leave out the serious considerations of the philosophy of science and ethics, cut down the long threads of prior research into just a few key references, simplify carefully designed methods and thoughtful data analysis into a few simple figures and bullet points, and often give up the idea of a thoughtful, cautious discussion in favor of a simpler set of conclusions. Is the whole philosophical heritage of science being reduced to a mere advertisement?

Although this may seem like something of a letdown, it is in some ways a sign of how successful science has become. Natural philosophy used to be mostly a hobby for the rich, practiced by just a few dozen people around the globe. Now there are hundreds of thousands of people doing science as a source of their livelihood. Governments, industry, and academia all find it worthwhile to pay people to do science. People whose ancestors were farmworkers, miners, laborers, and shopkeepers are now scientists. But this also means that thousands of students compete for the best jobs as scientists and very often do research as part of their training. They present this research at conferences, where the number of researchers requires that many scientists can only show a poster of their research, and these posters have to grab the attention of other scientists who are walking by dozens or even hundreds of posters. That is how we come to having posters that might be thought of as advertisements of research.

Does that mean that the lofty goals of understanding nature are gone? Does it mean that science has just become a competition for jobs, promotions, grant money, and fame, without any regard for ethics or the noble search for truth? I am sorry to say that a large number of scientists feel that way, at least on occasion. But I am convinced that being a scientist isn't just about getting a job. It is still about being a good human who uses the knowledge, methods, and skills of generations of scientists to learn more about the world. It is essential to the practice of science and our ethical obligations as humans to communicate that knowledge to other scientists and other humans. Sometimes, that communication will be little more than a poster, a press release, or even a tweet—but those brief communications have to be backed up by all the honesty, care, and thoughtfulness that have been the foundations of natural philosophy and science from the very beginning.

That's why I didn't want to write a book like the ones that some people have written—ones with titles like *How to Succeed in Science*. It's why I started with a lot of history and philosophy. Yes, self-promotion has become a part of scientific careers, just as it is part of the careers of salespeople and politicians. We need to remember, though, that one reason that academia, government, and industry want to pay scientists to work for them is because they are good at discovering the truth about the world. The only way they can do that is by behaving as good, honest scientists.

In the first chapter, I suggested that the qualities we tend to associate with scientists are inherently human qualities—being social, being communicative, wanting to explain the world, and being good at solving puzzles. What we learned in the rest of chapter 1, and throughout the rest of the book, was that the scientific community has evolved social structures and communication protocols that enable them to collectively come up with explanations of the world and solutions to puzzles that are more reliable than any one person could possibly produce. They are able to do this only because they value communalism, universalism, disinterestedness, originality, and skepticism, and because they try to be honest with themselves and others as much as possible. Advertising and promotion can sometimes give you a temporary advantage in the world, but to really succeed as a scientist over the long term, you will find that devotion to these more traditional scientific values is much more effective than a well-designed poster presentation.

People major in science for a variety of reasons. Some may have been convinced to major in science because they think it will improve their prospects for getting a good job. Others may be motivated by the possibility of finding technological or medical solutions to problems facing humans. Still others may just love nature, or may just enjoy the process of finding things out or solving puzzles. Regardless of why people might have entered science, however, through scientific training they become connected to something that is larger than themselves, and that has utility far beyond their initial objectives. It is indeed a thrill when you find that things you learned in separate classes and separate disciplines come together in a heightened consciousness of how the universe works, and discover how you can use that awareness to make predictions about the world that turn out to be true.

But don't let such successes go to your head! Remember, too, what we learned from Hume and Popper: that none of our understandings are infallibly, deductively correct; our present understandings are just tentative, and parts of our theories may have to be modified or abandoned. Even without the abstractions of philosophy, though, there are simpler reasons to be humble: people make mistakes, and scientists are people! So, remember why organized skepticism is part of the scientific ethos and the publication process. Try to be self-critical and bend-over-backward honest. Go ahead and make a poster that a promotes your research, but hope that someone visits your poster and asks questions that are challenging and make you think in new ways.

NOTES

1. What Does It Mean to Be a Scientist?

1 Popper, K. R. *The Logic of Scientific Discovery*, 91 (Routledge, 2002).
2 Latour, B. The science wars: a dialogue. *Common Knowl.* **8**, 71–79 (2002).
3 Latour, B. Why has critique run out of steam? From matters of fact to matters of concern. *Crit. Inq.* **30**, 225–248 (2004).

2. What Should We Do, and Why? The Questions of Ethics

1 Sen, A. *Human Rights and Asian Values* (Carnegie Council on Ethics and International Affairs, 1997).
2 Aristotle. *Aristotle's Nicomachean Ethics*, trans. Bartlett, R. C. & Collins, S. D. (University of Chicago Press, 2012).
3 Feynman, R. P., Leighton, R. & Hutchings, E. *"Surely You're Joking, Mr. Feynman!" Adventures of a Curious Character* (W. W. Norton, 1997).
4 Merton, R. K. Science and the social order. *Philos. Sci.* **5**, 321–337 (1938); and Merton, R. K. A note on science and democracy. *J. Leg. Polit. Sociol.* **1**, 115–126 (1942); both reprinted in Merton, R. K. *The Sociology of Science. Theoretical and Empirical Investigations* (University of Chicago Press, 1974).
5 Ziman, J. M. *Real Science. What It Is, and What It Means* (Cambridge University Press, 2000).
6 Miki, Y. *et al.* A strong candidate for the breast and ovarian cancer susceptibility gene BRCA1. *Science* **266**, 66–71 (1994).
7 Taylor, E. C. *et al.* A dideazatetrahydrofolate analog lacking a chiral center at C-6: N-[4-[2-(2-amino-3,4-dihydro-4-oxo-7H-pyrrolo[2,3-d]pyrimidin-5yl)ethyl[benzoyl]-L-glutamic acid is an inhibitor of thymidylate synthase. *J. Med. Chem.* **35**, 4450–4454 (1992).

8 Merton, R. K. A note on science and democracy. *J. Leg. Polit. Sociol.* **1**, 115–126 (1942). Page 123.

9 Here again, we must note that Merton's original formulations don't really fit closely to our current awareness. "The virtual absence of fraud in the annals of science" is one fact he attempts to explain within his discussion of disinterestedness. In the twenty-first century we are all too aware that fraud in science is more common than we would like, and some would contend that it has always been with us. See, for example, Broad, W. J. & Wade, N. *Betrayers of the Truth* (Simon and Schuster, 1983).

10 Merton's original norms formed the initialism "CUDOS," similar to the Greek word *kudos*, meaning praise, with the "OS" from "organized skepticism." Ziman substituted "originality" and "skepticism" and maintained the same initialism, while adding another norm.

11 When the Royal Society was formed in 1662, it received a charter from England's King Charles II, so government was clearly involved at that point. A perusal of the first volumes of the oldest scientific journal, *Philosophical Transactions of the Royal Society*, shows that a lot of what the members of the Royal Society were interested in had to do with commerce and industry. So even in the seventeenth century, the interests of government, industry, and science overlapped. In France, the Paris Academy of Sciences was even more closely tied to government and was assigned the role of advising the French monarchy on all sorts of technical and scientific matters.

3. The Scientific Literature: An Overview of the Terrain, and a Brief Hike In

1 Robert K. Merton—the same sociologist Merton we met in chapter 2—has researched the "shoulders of giants" metaphor and has found that it was not original with Newton but probably originated with Bernard of Chartres in the twelfth century. By the time of Newton, it seemed to be a fairly common saying. Merton, R. K. *On the Shoulders of Giants. A Shandean Postscript* (University of Chicago Press, 1993).

2 Giles, J. Internet encyclopaedias go head to head. *Nature* **438**, 900–901 (2005).

4. Scientific Journals, Past and Present

1 For example, Emiliani, C. *Planet Earth. Cosmology, Geology, and the Evolution of Life and Environment* (Cambridge University Press, 1992).

2 Grant, E. *The Foundations of Modern Science in the Middle Ages. Their Religious, Institutional, and Intellectual Contexts* (Cambridge University Press, 1996).

3 The Earl of Sandwich. An account of some observations, lately made in Spain, by His Excellency the Earl of Sandwich. *Philos. Trans.* **1**, 390–391 (1665). Note that the author of this contribution is the Earl of Sandwich,

who is employed as an ambassador to the King of Spain. At this time, there were few professional scientists; the people who did science were generally well-off educated gentlemen who engaged in science as a hobby.

4 An accompt of some mineral observations touching the mines of Cornwal and Devon; wherein is described the art of trayning a load; the art and manner of digging the ore; and the way of dressing and of blowing tin: communicated by an inquisitive person, that was much conversant in those mines. *Philos. Trans. R. Soc. Lond.* **6**, 2096–2113 (1671).

5 Gross, A. G., Harmon, J. E. & Reidy, M. S. *Communicating Science. The Scientific Article from the 17th Century to the Present* (Parlor Press, Inc., 2001).

6. Using Cited References—Backward and Forward

1 Gross, A. G., Harmon, J. E. & Reidy, M. S. *Communicating Science. The Scientific Article from the 17th Century to the Present*, 180. (Parlor Press, Inc., 2001).

2 Wang, H. & Sung, S.-S. Molecular dynamics simulations of three-strand β-sheet folding. *J. Am. Chem. Soc.* **122**, 1999–2009 (2000).

3 Barkley, D. *et al.* The rise of fully turbulent flow. *Nature* **526**, 550 (2015).

4 Snow, E. T., Foote, R. S. & Mitra, S. Kinetics of incorporation of O6-methyldeoxyguanosine monophosphate during in vitro DNA synthesis. *Biochemistry* **23**, 4289–4294 (1984).

5 Wolfe, J. P., Singer, R. A., Yang, B. H. & Buchwald, S. L. Highly active palladium catalysts for Suzuki coupling reactions. *J. Am. Chem. Soc.* **121**, 9550–9561 (1999).

6 Garfield, E. Citation indexes for science: a new dimension in documentation through association of ideas. *Science* **122**, 108–111 (1955).

7 Falagas, M. E., Pitsouni, E. I., Malietzis, G. A. & Pappas, G. Comparison of PubMed, Scopus, Web of Science, and Google Scholar: strengths and weaknesses. *FASEB J.* **22**, 338–342 (2008).

7. Reading a Scientific Paper

1 If you think scientists shouldn't just look at pictures, just read this little anecdote by Nobel Prize-winning physicist Richard Feynman in describing his rediscovery of a paper he had read years earlier: "I went out and found the original article on the experiment that said the neutron-proton coupling is T, and I was shocked by something. I remembered reading that article once before (back in the days when I read every article in the Physical Review—it was small enough). And I remembered, when I saw this article again, looking at that curve and thinking, 'That doesn't prove anything!'" It wasn't the words or arguments of the paper that he remembered—it was the graph! From Feynman, R. P., Leighton, R. & Hutchings, E. *"Surely You're Joking, Mr. Feynman!" Adventures of a Curious Character*, 233 (W.W. Norton, 1985).

8. Peer Review

1 This sort of review-by-member system, in a somewhat altered form, prevailed in the United States National Academy of Sciences through much of the twentieth century. By the end of the century, however, reforms to the system meant that submissions to the Proceedings of the National Academy of Sciences (USA) went through a process more like peer review in other journals.
2 Godlee, F., Gale, C. R. & Martyn, C. N. Effect on the quality of peer review of blinding reviewers and asking them to sign their reports: a randomized controlled trial. *JAMA* **280**, 237–240 (1998).
3 Godlee, F. & Dickersin, K. Bias, subjectivity, chance and conflict of interest in editorial decisions. In *Peer Review in Health Sciences*, 91–117 (BMJ Books, 2003).
4 van Rooyen, S., Godlee, F., Evans, S., Smith, R. & Black, N. Effect of blinding and unmasking on the quality of peer review: a randomized trial. *JAMA* **280**, 234–237 (1998).
5 Fletcher, R. H. & Fletcher, S. W. The effectiveness of journal peer review. In *Peer Review in Health Sciences*, 62–75 (BMJ Books, 2003).
6 van Rooyen, S., Godlee, F., Evans, S., Black, N. & Smith, R. Effect of open peer review on quality of reviews and on reviewers' recommendations: a randomised trial. *BMJ* **318**, 23–27 (1999).
7 Smith, R. Opening up *BMJ* peer review: a beginning that should lead to complete transparency. *BMJ* **318**, 4–5 (1999).
8 Akst, J. I hate your paper. *The Scientist* **24**, 36–41 (2010).

10. Refining Research Ideas and Writing a Proposal

1 Gross, A. G., Harmon, J. E. & Reidy, M. S. *Communicating Science. The Scientific Article from the 17th Century to the Present* (Parlor Press, Inc., 2001).

11. The Laboratory Notebook

1 Holmes, F. L. *Antoine Lavoisier—The Next Crucial Year, or, the Sources of His Quantitative Method in Chemistry* (Princeton University Press, 1998).
2 Kanare, H. M. *Writing the Laboratory Notebook* (American Chemical Society, 1985).

12. Scientific Writing: Grammar and Style

1 Hofmann, A. H. *Scientific Writing and Communication. Papers, Proposals, and Presentations* (Oxford University Press, 2016).

13. Assembling and Writing a Scientific Paper

1 If you've read chapter 1 carefully, you'll recognize that metascientists will question whether it is really possible to have results without some interpretation, and whether there is a clean distinction between facts and interpretations. In the case of separating out the *Results* and *Discussion*, however, the existence of what philosophically might be considered interpretation in the *Results* need not be a complication. The interpretations in the *Results* are likely to only be interpretations that are already well accepted and uncontroversial and require no explanation. The interpretations in the *Discussion* will be less obvious, potentially controversial, and require explicit explanation and argument.

2 If you have access to older volumes of the Proceedings of the National Academy of Sciences (USA), the Information for Authors in the mid-1990s had a lot of good examples of how to prepare figures. The Information for Authors for most journals was typically found in the first issue of each volume at this time; nowadays, it can also be found on a journal's website.

3 Büchner, E. Filtration vermittelst des Dr. R. Hirsch'schen Patent-Trichters. *Chem. Ztg.* **12**, 1277 (1888).

4 It would be nice if the *y*-axis in figure 13.1 were in the same units as the residual intensity in table 13.1, so we could better compare them. Ideally, we would go back and redo the experiment whose results are in table 13.1, using the more accurate quantitative method that was developed to obtain the quantitative results shown in figure 13.1.

5 Moray, R. A relation concerning barnacles, by Sr. Robert Moray, lately one of His Majesties council for the Kingdom of Scotland. *Philos. Trans. R. Soc. Lond.* **12**, 925–927 (1677).

6 Yorisue, T., Hayashi, R. & Ikeguchi, S. Distribution and orientation patterns of the pedunculate barnacle *Conchoderma* sp. on the swimming crab *Portunus trituberculatus* (Miers, 1876). *Crustaceana* **89**, 383–389 (2016).

7 Rosner, J. L. Reflections of science as a product. *Nature* **345**, 108 (1990).

8 Aronson, J. When I use a word ... declarative titles. *QJM* **103**, 207–209 (2010).

9 Grabocka, E. & Bar-Sagi, D. Mutant KRAS enhances tumor cell fitness by upregulating stress granules. *Cell* **167**, 1803–1813 (2016).

10 Rosner, J. L. Reflections of science as a product. *Nature* **345**, 108 (1990).

11 See preceding note. Rosner, Reflections of science as a product.

12 Aronson, J. When I use a word ... declarative titles. *QJM* **103**, 207–209 (2010); Pringle, J. R. An enduring enthusiasm for academic science, but with concerns. *Mol. Biol. Cell* **24**, 3281–3284 (2013); and Day, R. A. *How to Write and Publish a Scientific Paper* (Oryx Press, 1994).

13 Méndez, D., Ángeles Alcaraz, M. & Salager-Meyer, F. Titles in English-medium astrophysics research articles. *Scientometrics* **98**, 2331–2351 (2014).

14. Oral and Poster Presentations

1   Garner, J. K. & Alley, M. P. How the design of presentation slides affects audience comprehension: a case for the assertion-evidence approach. *Int. J. Eng. Educ.* **29**, 1564–1579 (2013).
2   See preceding note. Garner & Alley, How the design of presentation slides affects audience comprehension, p. 1576.

INDEX

Page numbers in bold refer to tables. Page numbers in italics refer to figures.